Elvira Powers

Hospital pencillings

Elvira Powers

Hospital pencillings

ISBN/EAN: 9783337124717

Printed in Europe, USA, Canada, Australia, Japan

Cover: Foto ©berggeist007 / pixelio.de

More available books at **www.hansebooks.com**

HOSPITAL PENCILLINGS;

BEING

A DIARY

WHILE IN JEFFERSON GENERAL HOSPITAL,
JEFFERSONVILLE, IND., AND OTHERS AT NASHVILLE
TENNESSEE,

AS

MATRON AND VISITOR.

By ELVIRA J. POWERS.

—— " And at each step,
His bloody falchion makes
Terrible vistas, through which victory breaks."

" We may tread the sick-bed floors
Where strong men pine,
And, down the groaning corridors,
Pour freely from our liberal stores
The oil and wine."

BOSTON:
EDWARD L. MITCHELL, 24 CONGRESS STREET.
1866.

TO OUR

LOYAL COUNTRYWOMEN,

WHO, AT HOME, IN CAMP, FIELD, OR CITY HOSPITAL,

BY PURITY, TRUE WOMANLINESS, AND THAT LABOR WHICH IS

WORSHIP,

"BY EVERY DEED OF GOOD THAT ANYWHERE
MAKETH THE HANDS OF HOLY WOMAN WHITE;"

HAVE CONTRIBUTED TO THE AID, COMFORT AND CONSOLATION

OF OUR

SICK, WOUNDED, AND DYING

"BRAVES IN BLUE;"

AND TO THE MEMORY OF THOSE WHO HAVE FALLEN VICTIMS

TO THIS LABOR OF LOVE,

IS THIS HUMBLE VOLUME

AFFECTIONATELY DEDICATED.

PREFACE.

SOME one has said "books are our best friends." If this be true, what need is there to offer an apology for presenting one more to the notice of the public? And yet, as the assertion is true only of those which leave no stain upon the mind, but rather like the insensible action of the sun and dew upon the bleaching linen on the greensward, leave it purer than before, it is well to know what is presented for our perusal.

The inducement which the author had for offering this little volume to the public was the fact that at the commencement of this four years' war, it was next to impossible to obtain any information upon the subject of the duties and trials of women nurses in military hospitals. This fact, together with the terrible stories afloat with regard to such duties and trials, was the cause of her not entering the service two years sooner. The papers were almost silent upon the subject until about this time, when the little which appeared was read with exceeding avidity.

Although this war has eliminated much in the way of musical, poetical and literary talent, yet comparatively little has appeared pertaining to the minutia of hospital service. But that little has such a charm for the author, that she has hoped her own record of experiences and observations may be received with something of the feeling with which she welcomes others.

Indeed she has some earnest hope of this in the fact that some part of " Hospital Pencillings " were published in the *New Covenant,* of Chicago, Illinois, where they " attracted much attention."

Not professing to treat of hospital life as a whole throughout the country, it gives a simple record of scenes and events, just as they occurred from day to day under her own observation. Thus much for the matter.

As regards the manner, if it will have any influence towards softening the point of the critic's stiletto to know that much of it was written in a sick ward, while surrounded by sick and wounded soldiers, momentarily interrupted with questions, or in short intervals of leisure between caring for their wants, or after the day's labor was over and the worn nerves called instead for

" Tired nature's sweet restorer,"

he is more than welcome to a knowledge of the fact.

In conclusion, if this little volume will serve to awaken a deeper interest in and a wider appreciation of our invalid and crippled soldiers, as

" They are coming from the wars,
And bringing home their scars,"

so that they shall be benefitted by such interest and appreciation, one more cause for gratitude to the All-Father will be granted

THE AUTHOR.

CONTENTS.

HOSPITAL PENCILLINGS.

CHAPTER I.

A TRIP TO DIXIE.

" How they went forth to die!
Pale, earnest thousands from the dizzy mills,
And sunburnt thousands from the harvest hills,
Quick, eager thousands from the city's streets,
And storm-tried thousands from the fisher's fleets,
How they went forth to die!

How ye went forth to save!
O Merciful! with swift and tireless heed
Along the myriad ways of pain and need,
With laden hand and ever watchful eye,
Fixed on the thousands going forth to die!
How ye went forth to save!"

ON BOARD THE " GEN. BUELL,"
OHIO RIVER, *April* 1, 1864.

HAVING been duly commissioned and ordered to " report
immediately at Nashville, Tenn., for hospital service at the
front," my friend, Miss N—— O——, and myself find
ourselves steaming down the Ohio, between Cincinnati and
Louisville.

Thus far we are quite ignorant of the duties of hospital
life, though so soon to enter upon them. Our Northern friends
have been questioned to little purpose, except that of ascer-
taining how very little knowledge there is upon the subject;
and the papers are equally silent.

1

This fact determines me to keep some sort of a journal, however imperfect. It will of course necessarily be so, as I must neglect no duty for the sake of scribbling about it.

We have just been seeking information of our gentlemanly escort, Mr. R., of Louisville. He, it appears, has an innate love of humor and a peculiarly dry and quiet way of quizzing people. Here was a fine opportunity. But we determine to ward off the attacks as skilfully as possible with the little knowledge we do possess. He says :—

"Well, ladies, I suppose you are prepared to make bread and gruel, sweep and mop, make beds, dress wounds and *plough ?* "

In reply the gentleman was informed that had we not been proficient in each, especially the ploughing, we should never have dared to make application for the situation.

He explained by informing us that one of the Southern refugees, who confessed herself unable to do either of the others, said she "could plough."

"And I suppose you have each brought good knives along with you?" was the next query.

"Knives—oh yes, but for what purpose do you mean?" And visions of being set to amputate limbs or to protect ourselves against personal assaults flitted through our minds.

"Well, nothing, only you'll have an enormous amount of onions to peel for those boys down there. You can peel those during the night, for you'll hardly have time in the day, that's the way I used to do."

"Did you? That's pleasant employment. I've practised it considerably myself, but didn't, like you, have the satisfaction of knowing during the grievous operation that I was shedding tears for the good of my country."

Then he wished to know whether in our visits to the sick wards we should "notice only the good looking ones." Upon being informed that we have fully determined to minister to such only as looked as if they were ministers, doctors, lawyers or editors, the gentleman seemed satisfied that we were fully fitted for the service. Still he felt called upon to caution us against excessive attention even to such, by relating that one of the class was asked by a lady visitor if she might " comb his hair."

"Yes—you—*may*," meekly responded the sufferer, " but it will be the thirteenth time to day."

Evening.

Just at sunset we passed North Bend, and had a glimpse of the tomb of President Harrison. The remains of Mrs. Harrison have within the last thirty days been laid by the side of the old hero. The place was pointed out by Dr. S., of Louisville, who is a second cousin to Mrs. Harrison. He informed us that the brother of his grandfather received a grant of all the land lying between the "Big and Little Miami," and extending back sixteen miles from their mouths. 4500 acres of this was willed to the grandfather of the Doctor and about the same to the mother of Mrs. H.

Dr. S. also informed us that he was the only one in Louisville who voted for Lincoln. That the polls were twice declared closed, and the clerk with oaths refused to record his vote, when the son of one of our Generals—I regret having forgotten the name—peremptorily ordered it done; when an A. and L. and a long black stroke was dashed upon the record. The baser sort had all day threatened hanging him upon the back porch, but at the close of the day most of them were safely intoxicated.

The Doctor has the sad trial of losing a son, who had by the offer of military emolument been drawn into the Confederate service. He was wounded or taken sick and carried to

Ohio, where a brother took care of him till his death. The father wished him brought home, and funeral services performed, but the military authorities of Louisville forbade it, as similar occasions had drawn out crowds of two or three thousands of secession proclivities. Then he was buried in Ohio, but when the citizens of the loyal little town learned that he had been in the Confederate service, they obliged Dr. S. to remove the body. That such staunch loyalists should suffer innocently is one of the saddest features of this rebellion.

In the course of conversation this evening we were informed by the Doctor that we were to pass the next day within seven miles of Mammoth Cave. And he spoke of the subterranean streams and mills in the vicinity, and of the blind fishes in the waters of the Cave.

"Yes," said Mr. R., in his usual serious way, "and I believe that is where your people go a craw-fishing!"

The Doctor replied in the affirmative, but in a tone which excited my curiosity. Here was a chance to add to my rather meagre stock of knowledge in natural history, and with the anxiety of a reporter for something out of which to manufacture an item, I inquired what kind of fish those were—if that was the name given to those blind fishes in the cave. To my astonishment a universal laugh greeted me from the trio. An explanation followed; and it seems that the same or something similar to what at the North we find in creeks and ditches, and call fresh-water crabs, there bear the name of craw-fish. And moreover as those crawl backward, they have attached a meaning to the term, so that when a man "puts his hand to the plough and looks back," he is said to have "gone a craw-fishing." So, like that notable traveller in Pickwick Papers, I can make a note of the discovery of a new kind of fish of the skedaddle genus. Hallicarnassus was decidedly

wrong in thinking one can sail around the world in an arm-
chair. He should have considerately assisted that big trunk
down stairs, and benignly seconded Gail's efforts to go abroad
and see the world, for peradventure she might learn something
even about craw-fish.

SATURDAY, *April* 2.

Reached the " City of the Falls " in the night. Left the
boat about six this morning, took a hasty breakfast at the
" National," then a hack for the depot, calling at the office of
Provost Marshal to secure passes on train to Nashville. Am
pleasantly impressed with Louisville. A pretty green plot
in front of private residences, even if quite small, with linden,
ailanthus and magnolia trees, are peculiarities of the city. It
is too early for the foliage of the trees to be seen, but the
deep green, thick grass and the blossoms of the daffodil are
in striking contrast to the snow I saw in the latitude of Chicago and Buffalo only day before yesterday.

The cars are now so crowded with soldiers *en route* for
" the front," that it is quite difficult for citizens to find passage. Some have to wait several days before they can find an
opportunity. Only one car is appropriated for this use, and
ladies with their escort always have the preference. Thus
gentlemen who are alone are liable to be left. As we were
leaving the " National " this morning a gentleman rushed out
and inquired if we were going to take the Southern train,
and if there was only one gentleman to the two ladies. He
" begged pardon—knew he was a stranger—wished to go to
Bowling Green—his wife was sick and he had written her he
would be home to-day. If the ladies would be so kind as to
pass him along, and if the gentleman would step with him
into the office he could convince him, through the keeper of
the " National," that he was a man of honor,"

2

Mr. R. referred the matter to the ladies. They decided to take under their protecting wing the lone gentleman and see him safe home if the interview with the landlord, with whom Mr. R. was fortunately acquainted, should prove satisfactory. It was so, and Mr. Moseby—not the guerilla as himself informed us—entered the hack. He had "taken the oath of allegiance," he said, and "lived up to it, but had a right to his own thoughts."

Upon arriving at the depot found the ladies' car locked, and we were left standing by it while the two gentleman looked after the baggage. Mr. R. was not to accompany us farther. Soon an elderly, pale-looking man, with a white neck-tie, came up, who asked if we each had a gentleman travelling with us. We hesitated and evaded the question. This was being in too great demand altogether. It was not even included in Mr. R.'s list of our duties. He "was really hoping we had not, and that one of us would take pity on an old man and pass him along."

His fatherly look and manner banished selfishness, and he was told to wait until the gentlemen returned, and we would see about it. As they did so Mr. Moseby stepped up and cordially shook hands with the old man, calling him "Judge." But all Southerners are styled judges, captains, colonels or generals, thought I, and this one is an honest old farmer nevertheless. As Mr. M. assured us that he was "all right," and a "man of honor," I told him he might occupy half of my seat in the car. But it was not long before I found that my poor old farmer was no less a personage than Judge Joseph R. Underwood, one of the most noted men and pioneers of Kentucky. He has been Judge of the Supreme Court of that State six years, a United States Representative for ten years and a Senator for six.

A spruce little Captain came through to examine military

passes before the cars started. Quite a number of citizens were left as usual, and as we were moving off I heard one young man exclaim in desperation that he would " go right back to the city and marry." The gentlemen congratulated themselves upon their good fortune, and the subject elicited the following incidents:

A gentleman of Mr. M.'s acquaintance could get no admission to the cars, no lady would take him under her care, and he asked the baggage agent if he might get in the baggage car. That functionary said he had orders to admit no one.

" Then you'll not give me permission, but if I get in will you put me out ? "

No answer was made, but the agent walked away, and the man, thinking like children, that " silence gives consent," entered the baggage car and remained.

Another gentleman, a merchant of Bowling Green, by name F— C—, could get no chance to ride. But fortunately having on a blue coat, in desperation he stepped up to a man with the two bars on his shoulder who was putting his soldiers aboard, and said with a pleading look and tone:

" Captain, can't you lengthen out my furlough just two days longer ? "

" No," said the Captain, in a quick authoritative tone, " you've been loafing 'round these streets long enough, in with you," and he made a motion as if he would materially assist his entrance if he didn't hurry.

" Well, if I must I must, but its *hard*, Captain."

" No more words," was the short reply, " in with you."

Another was related by an eye witness. A lady who was travelling alone was about stepping into the car, when a gentleman, who was trembling with anxiety lest he should be left, stepped up and offered to take her box. He did so, and stepping in behind was allowed a seat by her side, cautiously

retaining the box.   He had two comrades equally desirous of
securing a passage, who had seen his success.   One of them
stepped to the car window and whispered him to pass out the
box.   It was slyly done, and the gentleman marched solemnly
in with the weighty responsibility.   The box went through
the window again, and again walked in at the door, until it
must have been thoroughly " taken in " as well as the guard.

Just out of the city we passed a camp and saw soldiers
lying under the little low " dog tents " as they are called, and
in the *deep, clay mud*, while only a few rods distant was a
plenty of green sward.   Any officer who would compel his
men to pitch tents where those were ought to be levelled to
the ranks.

I saw for the first time to-day, fortifications. stockades, rifle-
pits, and mounted cannon at the bridges.   We passed over the
battle-ground of Mumfordsville, and saw the burnt fences and
the levelled trees which were to obstruct the march of our
troops, and the building which was used by them as a hospital.   In the deep cut passes one sees suddenly the picturesque
figure of a negro soldier, far above upon the heights, who
with shining uniform and glittering bayonet stands like a
statue, guarding the portals of liberty.   At the fortifications
are sign-boards upon which are printed in large letters,
" Please a drop a paper," while perhaps half a dozen hands
point to it as the train whirls past.   Some papers were thrown
out.   There were other things which had for our Northern
eyes the charm of novelty.   A half respectable or squalid
farm-house, with a huge chimney upon the outside, and with
a huddle of negro quarters.   Also negro women with turbans
upon their heads, working out of doors, and driving teams—
in one case on a load of tobacco, while driving a yoke of
oxen.   The total absence of country school-houses, and the
squalid and shiftless appearance of the buildings and people

at the depots, are in striking contrast to the neat little towns of the Northern and Eastern States. The scenery is fine, much of the soil good, and the water-power extensive. Nature has dealt bountifully with Tennessee and Kentucky, but the accursed system of slavery has blasted and desolated the land, and both races, black and white, are reaping the mildewed harvest.

I find my honorable companion very entertaining and instructive. I am indebted to him for many items of interest, both concerning the early settlers, and also the modern history of the places we pass. His personal history is full of interest, and is one more proof that early poverty is not necessarily a barrier to honor and position. The Judge was given away by his parents to an uncle, who educated him, gave him five dollars and told him he must then make his own way in the world. Another uncle lent him a horse, and he set out to seek his fortune as lawyer and politician. He has in trust the fortune of an eccentric old bachelor, which is known in Warren County as the Craddock fund. Three-fourths of this is used to educate charity children, while the other fourth pays the Judge for his care of the fund. His friend Captain C., while upon his death-bed, sent for the drummer and fifer to play tunes in the yard, and from those selected such as he wished played at his funeral. He was buried with military honors.

" Muldroughs-Hill " which we saw, is a long ridge extending about one hundred miles from the mouth of Salt-River to the head of Rolling-Fork. It was named from an early settler who lived twenty miles from the others, and was farthest west. Rolling-Fork is a tributary of Salt-River. The origin of the term " going up Salt-River " originated at a little place we passed, now called Shepherdsville. It has only four or five hundred inhabitants. But in its early days its salt licks supplied all the Western country with salt, and was a growing aspirant for popularity, as it invited so much trade. It was a rival of Louisville, but unlike that, made no provision for its future well-being, but depended on its present worth alone. "Thus," moralized the Judge, "do we often see two young men start out with equal advantages, and find afterward that one became a Shepherdsville, and the other a Louisville." Now there is a bridge at Shepherdsville guarded by cannon, then there was no bridge and ferry-boats were used. It was not a smooth stream, and to cross, one must row up the river some one hundred rods before heading the boat to the opposite shore. Owing to the rapidity of the current, it was hard rowing, and great strength was needed. There were those engaged in the making of salt who were called kettle-tenders, and who for the most part were a low, rough set, being often intoxicated and quarrelsome. Two of these having a fight, the victor finished with the triumphant exclamation of

"There, I've rowed you up Salt River!"

Lincoln's birth-place is near this, in the adjoining County of Larue—although this was not the name at the time of his birth. And how little did the mother of Lincoln think, as she taught him the little she knew of books, that the people in the vicinity would ever have cause to exclaim of him, in relation to his rival for the Presidency, as they do of the successful politician—"he has rowed him up Salt River!"

There is a little river called "Nolin," which waters his birth-place. It was so named from the fact that in the early settlement upon its banks a man named Linn was lost in the woods, and never found. He was probably killed by the Indians. But the neighbors searched for several days, and at night met at a place upon its banks, calling to each other as they came in, "No Linn,"—"No Linn, yet."

The Judge has carried lead in his body for over fifty years, received in the war of 1812. He was in the battle on the Maumee river called Dudley's defeat. The regiment, under Dudley, had crossed the river to take cannon of the enemy, which they succeeded in doing, but instead of returning they pursued them two or three miles, leaving a few behind to protect the captures. But a detachment of the enemy passed around in their rear, retook the cannon, and when the regiment returned, their retreat was cut off, and all were taken prisoners and obliged to run the gauntlet. About forty were killed in running the gauntlet. The Judge saw that the line of men which had formed at a little distance from, and parallel with the river, had a bend in it, and that if he ran close to the guns they would not dare fire for fear of hitting their own men. The Indians were armed with guns, tomahawks, and war clubs. In that day the gun was accompanied with what was called the " wiping-stick," which was a rod made of hickory notched, and wound with tow, and used to clean the gun. He escaped by receiving a whipping with some of those sticks. It was the last gauntlet ever run in the United States.

During the trip I had quite a spirited but good-natured discussion upon the condition of the country, with Mr. M., who I found is really a strong rebel sympathizer. He worships Morgan since his late raid into Ohio, and secretly cherishes his picture in his vest pocket. Just before reaching Bowling Green, where we were to separate, the fatherly old Judge took a hand of each in his own, and with moisture in his eyes and a tremor in his voice, said :

" My children, you represent the two antagonistic positions of the country, and like those, do not rightly understand each other, on account of sectional prejudices. And now let an old man who has watched the growth of both sections, who has, as he trusts, fought for their good in the field, the desk, and senate, join your hands in the grasp of good fellowship, and oh, how sincerely I wish that I could bring also together the North and South in one lasting peace!"

Soon after, he pointed out his residence—the cars stopped, and we parted with our pleasant friends.

Reached the " City of the Rocks " about five, this P. M. Shall wait to see more of it, before making note of impressions.

CHAPTER II.

NASHVILLE, TENN., *Thursday Evening, April 7.*

The present week, thus far. has been to me, full of new and thrilling experiences.

On Sabbath, the day after our arrival, I entered an ambulance and visited a camp for the first time. The company consisted of three, besides myself—Rev. Dr. D., a young theological student who is passing vacation here, and Miss T. The day was warm and springlike ; the hyacinths, crocuses, and peach trees in blossom. It was the camp of the 7th Pennsylvania Cavalry. and situated upon one of the hights overlooking the City. The tents were white, the soldiers well-dressed, the uniform bright and everything tidy. A new and gaily painted banner pointed out the tent of the Colonel. As we entered the grounds, that gentleman, with the Major, met us cordially, a seat was prepared for the ladies at the opening of the Colonel's tent. while a huge box in front served for a speaker's stand. The bugle then summoned such as wished to listen. and service was held by the two gentlemen of our party. Books and papers were afterward distributed, for which the soldiers seemed eager. The Colonel informed us that the Regiment had just been reorganized, and new recruits filled the vacant places in the ranks, made so by the heroes, who fell at such battles as Lookout Mountain, Mission Ridge, and Chickamauga. There is a long list of such inscribed upon this banner, of which they are justly proud.

On Monday, visited a hospital for the first time. Was ac-

3

companied by Mrs. E. P. Smith, Mrs. Dr. F. and my travelling companion Miss O, beside the driver. As the ambulance halted, we saw through the open door and windows the home-sick, pallid faces raised from the sick beds to greet us with a look of pleasure. Upon entering, almost the first object was that of a dying boy. His name was John Camplin, of Co. G. 49th Illinois Vols. He was a new recruit of only seventeen, and the victim of measles. He "did'nt want to die," but, after the singing of such hymns as "Rock of Ages," and "Jesus lover of my soul," he grew more resigned. I took the card which hung in a little tin case at the head of his bed, and copied the name and address of his father. The dying boy had been watching, and he then with difficult speech asked me to write to his people and tell them "good bye," and that he was "going home." I tried to obtain a more lengthy message to comfort them, but speech was soon denied nd reason wandered. He died a few hours after, and the sad tidings was sent next day.

Found another poor boy quite low, with pneumonia. He knew his condition, but with an heroic smile upon his wasted features said, that "if" his "life would do his *dear* country any good" he was "willing to give it."

The Masonic Hall and First Presbyterian Church constitute Hospital, No. 8. We visited that on Tuesday.

As we enter the Hall, past the guard, we find a broad flight of stairs before us, and while ascending, perceive this caution inscribed upon the wall in evergreen.

"Remember you are in a hospital and make no noise." Up this flight, and other cautions meet us, such as "No smoking here"—"Keep away from the wall," &c. We here pause at a door, and are introduced to the matron who is fortunately just now going through the wards. It is Miss J—tt, of Ann Arbor, Michigan.

Ascending another broad flight, and asking in the meantime of her duties, she throws open the door of the linenroom where are two clerks, and says :

"This department comprises all the work assigned to me— whatever else I do is voluntary and gratuitous. "But today," she adds laughingly, "it would be difficult to define my duties. I think I might properly be called 'Commandant of the Black Squad,' or Chief of the Dirty Brigade;" and she explained by saying that she had seven negro women and two men, subject to her orders, who were cleaning the building. She next throws open the door of a ward which contains but a few patients, and has a smoky appearance. She tells us, they are fumigating it, having had some cases of small pox, most of which have been sent to the proper Hospital.

We pass to another, where she tells us, previous to entering, is one very sick boy. He is of a slight form, only fifteen, and with delicate girlish features. His disease is typhoid fever, from the effects of which he is now quite deaf. As we approach, he says to her faintly,

"Sit down here, mother, on the side of my bed."

She does so, when he asks her to " to bend her head down so he can tell her something." This she does, when he says, quite loud, but with difficulty ;—" There's some money under my pillow, I want you to get it, and buy me some dried peaches."

" I don't want your money," she says, " but you shall have the peaches if I can get them," and she writes a note and dispatches it to the sanitary rooms for them." " This boy always calls me mother," she says, "and the first day he was brought here, he sent his nurse to ask if I would come up and kiss him. He has always been his mother's pet, and I now correspond with her on his account."

His fever is very high, and we pass our cold hand sooth-
ingly over his forehead and essay to speak words of cheer,
and as we turn to leave, he looks up pleadingly and says :

" Can't *you* kiss me ?"

" Yes, indeed, I can—am glad to do so," and we press our
own to his burning lips and receive his feverish, unpleasant
breath, not a disagreeable task though, for all, when we remember that he is the pet of his mother, who misses him so
very much, and who may never look upon her boy again.

Of one—a middle-aged, despondent looking man we ask
cheerily. how he is to-day.

" About the same," he replies coldly, but with a look which
is the index of a thought like this :

" Oh, you don't care for us or our comfort,—you are well,
and have friends, and home, probably near you, and you cannot appreciate our suffering, and only come here to satisfy an
idle curiosity."

He does not say this, but he thinks it, and we read the
thought in the voice, manner, and countenance. We determine to convince him of his mistake, if possible, notwithstanding he looks as if he prefers we should walk along and
leave him alone.

" Were you wounded ? " we ask.

" No—sick," was the short gruff answer.

" Your disease was fever was'nt it ? " we persist,—" your
countenance looks like it."

" Yes, fever and pneumonia," he replies in the same cold,
but despairing tone.

" Ah—but you're getting better now."

" Don't know about it—reckon not."

" Well, how is it about getting letters from home ? "

His countenance, voice and manner undergo a sudden
change now, and his eyes overrun with tears, at the simple
words " letters from home."

And as he raises his hand to his mouth, to conceal its quivering, he tells us with tremulous voice that he has sent three letters to his wife and can get no answer. She has left the place where they used to live, and he does not know certainly where to direct. We ask who we can write to, to find out, and learn that a sister would know. We take the probable address of the wife, and that of the sister, and after some farther conversation leave him looking quite like another man as we promise to write to each in the evening. (Subsequently, we learned that he received a reply to both, and was comparatively cheerful and very grateful.)

Down stairs, and we enter a ward on the first floor. Here is a thin sallow visage, the owner of which piteously asks if we "have any oranges." "No," but we provide means, by which he can purchase.

"I'm from North Carolina," he says, "I hid in the woods and mountains and lived on roots and berries for weeks, before I could get away."

In reply to our query as to whether he would like a letter written home, he informs us that his wife and father arrived in town only a few days ago.

"Then you have seen them," we say.

"Yes, they both visit me, but my wife comes oftenest."

Just now, his nurse, a young man who should know better, interrupts him by telling us that "it isn't so, and his family are all in North Carolina."

"That's just the way," said the sick man, turning to me with a flushed and angry look. "that they're talking to me all the time, and trying to make everybody think I'm crazy. I reckon *I* know whether I've seen my wife or not!"

"Of course you do," we say quietingly; "does she bring you anything nice to eat?" and we add that we wish she would come while we were there, so we could see her.

4

"Well, she don't bring me much to eat," he says in a weak, hollow voice, but earnestly, "she don't understand fixin' up things nice for sick folks, and then she's weakly like, but she does all she can, for she's a right gude heart. She doesn't fix up, and look like you folks do, you know," he added, "for she's sort o' torn to pieces like by this war."

"Yes, we can understand it."

Upon inquiring about this man a few moments after of the Ward-Master, we find that he is really a monomaniac upon this subject, persisting in the declaration that his wife and father visit him often though no one sees them.

"He can't live," said the Ward-Master, "he has lost all heart and is worn out. The chance of a Southerner to live after going to a hospital is not over a fourth as good as for one of our Northern boys. They can do more fighting with less food while in the field, but when the excitement is over they lose heart and die."

We find upon several subsequent visits that he is growing weaker, and at the last when his countenance indicates that death is near, we are thankful that he is still comforted by these imaginary visits from father and wife.

We crossed the street and entered the First Presbyterian Church, which constitutes a part of the hospital. This place is notable for the promulgation of secession sentiments from its pulpit in other days. A specimen of the style was given here a short time before the entrance of our troops, by Prof. Elliott of the Seminary, who in a prayer besought the Almighty that he would so "prosper the arms of the Confederates and bring to naught the plans of the Federals, that every hill-top, plain and valley around Nashville should be *white with the bones of the hated Yankees!*"

After hearing this it was doubly a pleasure, in company with Miss J., another "Northern vandal," to make the walls of the old church echo to the words of "The Star Spangled Banner," with an accompaniment from the organ ; and it would have done any loyal heart good to see how much pleasure it gave to the sick and wounded soldiers.

SATURDAY EVE, *April* 9.

Last Wednesday Miss O. and myself visited Hospital No. 1, for the second time.

They were just robing one young boy in his soldier's suit of blue for the last time. He was then borne to the dead-house. His name was Hickman Nutter, of the 31st Ohio. I secured the Post Office address of his people and that of several others who had died and had no message sent home. I passed the whole of the next day in writing soldiers' letters, and in my journal. My fortitude was sorely tried and really broke down after getting back, to find that in ward 1 alone from two to four boys are dying daily, while the Chaplain has not been in to speak to a single sick or dying boy for two weeks. Wards 2 and 3 have fared little if any better, as is the testimony of ward-masters and nurses. It is his duty also to write to the relatives of those who die, and common humanity would dictate that it be done, and every comforting message sent to them. I was told by the clerk, whose duty it was to collect the names for report in the public prints, that in no single instance had he known the Chaplain to attend to that duty. I was indignant and determined to report him, but was given to understand by more than one Christian minister, that the expression of indignation was considered a bad omen for my future success in hospitals.

" People here," said one, kindly in explanation, " must learn to see and hear of all manner of evil and wickedness going on around them, and be as though they saw and heard not."

Being by nature and birth an outspoken New Englander, and having inhaled freedom of speech from the breezes which blow from the hills of the " Old Bay State," I fancy it will not be very easy becoming initiated into this phase of military service.

We found several interesting cases on passing through wards 1, 2 and 3.

In the first, saw one man in a dying condition, who was brought the night before. He was lifted from the ambulance and brought in by two men, who immediately left without being questioned or saying anything about him. The attendants were busy and expected to find all needed information in the medical papers, which it is rulable and customary to send, but which were not to be found. No one had observed the ambulance or men sufficiently to identify either. The disease could not be determined. There were no wounds and the lungs were in a healthy condition, but he was dying and insensible. A letter was fortunately found in his pocket, from his wife, which gave his name, company and regiment, as being Henry Clymer, Co. K., 128th Indiana.

In passing through ward 2 we came to a handsome young man, who was looking so well compared with others that we were passing without speaking. But the nurse said to us:

" This man is blind ! "

Could it be possible ! His eyes to a casual observer were perfectly good, but upon a closer examination one saw that the pupil was greatly enlarged and the expression staring and vacant. Questions revealed the fact that he could see nothing except a faint light when looking towards the window. I asked the cause.

" Medicine, the Surgeon here says," was the reply. " I had chills and fever while at the front, and the physician gave me

large quantities of quinine, which made me blind. I have the
ague now, but the Doctor dare not give any more quinine. I
have been blind two weeks."

" Doesn't the Surgeon think the medicine will leave your
system, and that you may recover your sight?"

" Well, he doesn't speak very encouragingly—says he
doesn't know."

And we now see that although the eyes cannot do duty in
one way they can in another, for they absolutely rain tears,
as he tells us with quivering lips, that his wife does not know
anything about it; that he is dreading to send her word by
stranger hands,—he cannot bear to think that may be he can
never write again,—never see her or other friends in this world.
He is yet young and life has looked so pleasant; he is a professing Christian, but finds it *so* hard to bear this affliction.
And he sobs like a whipped child, as, kneeling by the head
of his low bed, with hand upon his forehead, we listen to
this recital and strive to comfort him. We tell him of others
afflicted in the same way who have not passed a life of idle-
ness in consequence, but of mental or physical activity. Of
those who have risen superior even to this calamity, and in
the battle of life have learned

" How sublime a thing it is
    To suffer and grow strong."

He says our words have been a blessing, as we take his
hand in a good-bye, and with a promise to break the news to
his wife, as gently and hopefully as possible. [We do so
subsequently and upon the last visit find that he has been
gaining his sight so that he can distinguish forms, though not
features. Again we stand by his vacant bed and learn that
he with many others have been sent North to make room for

more sufferers from the front. But he was still gaining his sight.]

In the same ward we find one slight young boy, who looks as if he ought to be at home with his mother, and we sincerely believe is crying because he isn't—though he'd be bayonetted sooner than own it. He draws his sleeve across his red eyes as we approach, and upon our questioning informs us that he is " almost seventeen," and furthermore that he is " nearly half a head taller and two pounds heavier than another boy in his regiment;" but confesses that he is " right tired a' laying this way day after day—fact is I'd a heap sight rather be at home if I could get to go there, for I enlisted to *fight*, not to be *sick!* " Now we ask him if he ever thought while lying there that he is *suffering in the service of his country,* and a quick flash of the eye, a smile and an emphatic " no," tell us that it is entirely a new thought. Then we beg him not to forget that he is, and assure him that it requires a much braver soldier to suffer day after day in a hospital than on the hardest battle-field, and we leave him with a look of heroic endurance on his childish brow.

Here is a good-faced German, who is moaning with pain from an amputation. It is twenty days since the operation, but he suffers terribly every few moments from a spasmodic contraction of the muscles. And we also find upon conversing, that the fact of the amputation hurts his feelings in more ways than one, and we must needs tell him to bear the pain like a good brave soldier, and that it will grow less and less each day, and really last but a few days more altogether, and that as to being without a limb he will not be the only one capable of exhibiting such a proof of the service rendered his country, that it is an honor rather than a disgrace to lose limbs while battling for the right; and now the hero's look of determination settles over his features also. But just as we turn to leave, he expresses his opinion that two or three more such "cookies" as we brought him the other day wouldn't hurt him, indeed,

"Dey was mosht as goot rot my moder used to make."

SUNDAY EVE, *April* 10.

Attended church to-day at the Second Presbyterian, or "Union Church" as it is called. It is the only one in the city, I am told, where one is sure of hearing sentiments of loyalty. Rev. Mr. Allen is pastor. He does not fear now, under the shadow of Fort Negley, and with so many "blue coats" about, to "Lift up his voice like a trumpet, and show the people their transgressions and the house of Jacob their sins." I believe, however, that he was obliged to leave the place previous to the entrance of our troops.

I saw a pomegranate flower for the first time, to-day. It is of a dark red color, single, about the size of a plum blossom. It is of the same family I think, though cannot analyze it, for want of a botanical work.

In passing through ward 1 of the hospital last Wednesday, and asking advice of the chief nurse—who, by the by, is soon to complete his studies as surgeon—as to what we could do for the benefit of the invalids, he said there were two cases who would die unless some one could by attention and cheerful conversation save them. That they had been sick a long time, were very low, but the trouble now was nervous debility from homesickness and despair of life. Had himself done what he could for them, but was worn out with care of the ward and loss of sleep. And he added:—

"The Surgeon has given them up, and I will give them into your charge, and if they live it will be your care which saves them."

"Would anything be injurious for them to eat?"

"No, if you can get them to eat anything you will do better than I can."

Upon inquiring which they were, he pointed them out, when I told him that I had spoken to both only a few moments before, and that one would scarcely notice me enough to tell me his disease, while the other would not answer at all, but drew the sheet over his face.

"Oh, yes," he replied, "they think no one cares for them, that they're going to die, and the worst one is in a half stupor much of the time. But pass your hand gently over his forehead to arouse him, and then you know how to interest him."

He then directed the nurse of this one to go with me and see that everything was done which I directed. The nurse and patient were both from Indiana, and the former going to the side of the bed toward which the face of the sick man was turned, said in a peculiarly pleasant and sympathizing tone :—

"William, there's a lady come to see you and she wants to make you well if she can."

Passing my hand over his forehead, as directed, I added as cheerily as possible :—

"Yes, William, I've come to see if I can't do something for you ; if I shall write some letters for you, or bring you something to eat to make you better."

He roused up and I knew he was listening, but not wishing to excite him too much I then commenced asking of the nurse about his company and regiment, and the length of time he had been sick in that hospital. But I had scarcely done so, when the sick man turned his face down into the pillow, burst into tears and grieved and sobbed like a child, fairly shaking the bed with the violence of his emotion. The nurse bent down to him, and said as if pacifying a sick child :—

"Don't fret so, William, this lady loves you, and she's going to try to make you well."

I knew the tears would do him good, but I spoke low and slowly, and the sobs grew less as he listened :—

"You've been sick a long time, I know, and have grown discouraged and have thought you were never going to get well, but the Doctor says there is nothing to hinder if you will only try. I was once sick myself with a low nervous fever, and felt just as you do for a long time. And the physician told me at last that I wouldn't live unless I made up my mind to try to live. And I did try and worked hard for it for a long time else I should never have got well. And now if you will do the same and think all the time of what you are going to do when you get well, I will come and see you as often as I can, and bring you anything you wish to eat. Wouldn't you like to have me write for you to ask your wife, mother, or sister, to come and take care of you?

Just then the nurse tells me he is "single" and I repeat the question of his mother and sisters.

" No," he replied, in a sad, grieved, hollow voice, "*they wouldn't come.*"

"Shouldn't I write to his father to tell him how he was." " No," he didn't " want any letters written."

" Could he think of something he could eat."

He said he could not, but the nurse exclaimed :—" Why, William, don't you remember you said the other day you could eat some pickles, if you could get them?" " Yes, I could eat some *pickles*," said the slow, hollow voice. A little inquiry found that it was possible he could eat a cookie also, so it was arranged that the nurse should call at the home of the Christian Commission, where I was stopping, for the articles.

I also learned that the sick man had not been bathed since having the fever, and his face looked like dried parchment. I made a prescription of castile soap and warm water for his benefit, to be applied to the whole surface of his body—the

5

application to take place immediately after my departure. After the bath, the nurse called and I sent some cookies and a small jar of pickles.

The other patient to whom I was referred, was scarcely less interesting, but have not time to note the particulars. I visited them again yesterday, and found my directions with regard to each had been carried out, and both were better and glad this time to see me. William rejoiced in the jar of pickles upon his stand, out of which he had gained sufficient appetite to "reckon," he "could eat a few dried peaches, if he could get them." A small jar of those was prepared and sent to him, with a second edition of cookies.

TUESDAY, April 12.

Have visited Hospital. No. 8, as well as No. 1, several times since I have been here, and am priviledged to carry some delicacies, and write letters for its inmates.

I yesterday visited Hospital, No. 1, for the last time probably, while those remain in whom I have become specially interested. But have made such arrangements that William and the Alabamian, who were given to my care, shall have whatever is needed. They seem to regret my departure, but William is decidedly better. Carried a large bottle of lemonade, some oranges, and blackberry sirup.

Found a poor old Norwegian suffering terribly from the application of bromine to the gangrenous wound in his arm. He was very thankful for an orange and some lemonade—had eaten nothing for two days. His face and bald, venerable head were covered with a red silk handkerchief, to hide the great tears which were pressed out by the pain ; but his nurse said he never gave a word of complaint.

The German with amputated limb is easier—the blind man

hopeful of sight, and the little fellow improving, who "enlisted to fight, and not to be sick."

While in ward 3, yesterday, I was beckoned to, from a sick bed, whose occupant wished me to come and "rejoice with him." Upon going there he assured me with a mysterious air, that he "isn't going to tell everybody, but as I was a particular friend of his, and he had always thought *right smart* of me, he would tell me something greatly surprising."

Upon expressing my willingness to be surprised, he confidently and joyfully assured me that though very few people knew it, yet he was "*The veritable man who killed Jeff. Davis, President of the Confederate States!*"

He waited a moment to note the effect upon me of this pleasing intelligence, when I quietly told him I didn't know before that Jeff. Davis was dead, but that if he was, and he was the one who killed him, they ought to give him a discharge and let him go home, as he has done his share of the work. Then he joyfully assured me, that "they have promised to do so, and that his papers are to be made out to-morrow." But more serious thoughts came to me then, for I saw written upon his countenance, in unmistakable characters, the signature of the Death angel, marking his chosen, and though I knew not how soon his papers would be made out, was certain that before long they would be, and that he would receive a full and free discharge from all earthly toil and battle from the Great Medical Director of us all!

While passing through the aisles of wounded men, and hearing their stories, many of them intensely graphic, I seemed to hear something like the following, which, may the author whose name I do not know, pardon me for copying:

"Let me lie down,
Just here in the shade of this cannon-torn tree,—
Here, low on the trampled grass, where I may see

The surge of the combat ; and where I may hear
The glad cry of victory, cheer upon cheer :
          Let me lie down.

Oh, it was grand !
Like the tempest we charged, in the triumph to share :
The tempest—its fury and thunder were there ;
On, on, o'er intrenchments, o'er living and dead,
With the foe under foot, and our flag overhead,—
          Oh, it was grand !

Weary and faint,
Prone on the soldier's couch, ah ! how can I rest
With this shot-shattered head and sabre-pierced breast ?
Comrades, at roll-call, when I shall be sought,
Say I fought till I fell, and fell where I fought,
          Wounded and faint.

Oh, that last charge !
Right through the dread hell-fire of shrapnel and shell,—
Through without faltering, clear through with a yell,
Right in their midst, in the turmoil and gloom,
Like heroes we dashed at the mandate of doom !
          Oh, that last charge !

It was duty !
Some things are worthless, and some others so good,
That nations who buy them pay only in blood ;
For Freedom and Union each man owes his part :
And here I pay my share, all warm from my heart,
          It is duty !

Dying at last !
My mother, dear mother, with meek, tearful eye,
Farewell ! and God bless you for ever and aye !
Oh, that I now lay on your pillowing breast,
To breathe my last sigh on the bosom first prest !
          Dying at last !

I am no saint !
But, boys, say a prayer.   There's one that begins,—

' Our Father ; ' and then says, ' Forgive us our sins : '
Don't forget that part; say that strongly ; and then
I'll try to repeat it, and you'll say amen !
     Ah ! I'm no saint !

     Hark ! there's a shout !
Raise me up, comrades ! We have conquered, I know ;
Up, on my feet, with my face to the foe !
Ah ! there flies the flag, with its star spangles bright,—
The promise of Glory, the symbol of Right !
     Well may they shout !

     I'm mustered out !
O God of our fathers ! our freedom prolong,
And tread down rebellion, oppression, and wrong !
O land of earth's hopes ! on thy blood-reddened sod,
I die for the Nation, the Union, and God !
     I'm mustered out ! "

6

.

CHAPTER III.

NASHVILLE is a city which is set upon hills. It is also founded upon a rock, and the fact that it has not much earth upon that rock, is made the pretext for leaving numberless deceased horses and mules upon the surface, without even a a heathen burial, until they are numbered with the things that were.

But it has been comfortingly asserted by the agent of the Christian Commission here, Rev. F. P. Smith, that it is astonishing how much dead mule one may breathe, and yet survive.

Nashville is also a city of narrow, filthy streets, and in some localities, of water, which, like the "offence" of the king of Denmark, "smells to Heaven."

It is moreover a city of *mules*. Two, four, and six mule teams, with a driver astride of one of them, and sometimes with the high, comical-looking Tennessean wagons attached— not to the driver particularly, but to the mules. These, with *mulish* mules, who draw crowds instead of wagons, animate the streets day and night. It is a city of either dust or mud —but one street boasts a street-sprinkler.

The citizens of Nashville who remain, have mostly taken the oath of allegiance to protect their property, but it is esti- mated that not above one in fifty is, at heart, loyal. The ladies (?) sometimes show their contempt of Northern labor- ers by making up faces when meeting them upon the streets, but there are so many " blue coats " about, they do not think it advisable to allow their

" Angry passions rise,"
To tear out our eyes ;"

as they would evidently consider it a great pleasure to accom-
plish.

Nashville and its vicinity boasts a few distinguished per-
sonages beside myself. Mrs. Polk, widow of the Ex-President, resides a few blocks from this. Gen. Sherman's head-
quarters are at a lovely retreat, we think, on High Street,
and Gen. Rouseau's but a few blocks distant, while the Hermitage of Gen. Andrew Jackson is but twelve miles east of
the city. This has many visitors, but who seldom venture
now without a guard. Since our stay here, a party of four
ladies from Hospital, No. 19, with as many gentlemen, and a
guard of thirteen, visited the Hermitage, who learned next
day that a party of guerillas, 100 in number, came there an
hour after they had left, and followed them. At first, as they
informed us, they made it a subject for pleasant jesting, but
after farther consideration, for that of serious thought, as they
came rather too near being candidates for " Libby," or a
worse fate.

A nephew, who is also an adopted son of the old General,
has charge of the place; he has two sons in the rebel service. The property is confiscated to the Government, but the
family, out of respect to the memory of the stern old patriot,
are permitted to remain. The visitors may see here the
quaint and cumbrous family carriage in which the General
used to journey, together with a buggy, made from tho tim-
bers of the old ship Ironsides.

The family, especially the female portion of it, being.of
secession principles, keep themselves secluded from the gaze
of northern mudsills. But the mudsills, presuming upon the
cordial reception which they believe would be extended by
the General himself, usually make themselves sufficiently at

home to wander at their own sweet will through the grounds, and partake of a lunch on the shaded piazza.

It is a fine old mansion, approached by a circular avenue, which is shaded by grand old trees. And notwithstanding that the General has adopted grandsons in the rebel service, and his family are secessionists, yet it requires but little faith to believe that the stern old hero is not unmindful of the present gigantic struggle, neither a great flight of the imagination when the wind is moaning and stirring the lofty branches of the grand old trees, to fancy that his voice, in suppressed and now *reverent* accents, yet emphatically exclaims:—

"*By the Eternal, the Union must, and shall be preserved!*"

The city contains many elegant private residences, and splendid public buildings.

Among the latter is the State Asylum for the Insane, which has four hundred and fifty acres attached, and had an expenditure of $48,000 per annum. Another is the Institution for the Blind, the expenses of which for the year 1850, were nearly $8,000. The Tennessean Penitentiary is also a superior structure. In September 30, 1850, the number of inmates was three hundred and seventy-eight, and of this number three hundred and sixty-six, were white men, with only eight black men, three white women with only one black woman.

The Medical College is a fine building and contains a valuable museum. The University is an imposing edifice of gray marble, while the Masonic Hall, the Seminary and graded school buildings are spacious and beautiful structures. The *first* in importance, among the public buildings of Nashville, and which is second to none in the United States in point of solidity and durability, is the Capitol. This is a magnificent edifice, situated on an eminence one hundred and seventy-five feet above the river, and constructed inside and out, of a beautiful variety of fossilliferous limestone or Tennessee marble. At each end, it has an Ionic portico of eight columns, and each of the sides, a portico of six. A tower rises from the centre of the roof to the hight of two hundred and six feet from the ground. This has a quadrangular base surmounted by a circular cell, with eight fluted Corinthian columns, designed from the celebrated choragic monument of Lysicrates, at Athens.

Among the private residences we have seen, is a beautiful mansion, still unfinished, which, at the time of his death, was being built for the rebel Gen. Zollicoffer. A more unpretending one perhaps, is that of the widow of ex-President Polk, the grounds surrounding which contain his tomb—a plain, simple, temple-like fabric, of light brown marble.

That beautiful baronial domain known as the Achlen estate is situate about two miles out of town. For attractions it has extensive grounds, with great variety and profusion of shrubbery, among which flash out here and there, life-like statues of men and animals, and miniature monuments and temples. A fountain jets its diamond drops, while an artificial pond is the home of the tiny silver and gold fish. Beside the noble family mansion is another building nearly as spacious, which is used as a place of amusement. A well-filled conservatory is another beautiful feature, while an observatory, which crowns an imposing brick tower, gives a view of the scenery for miles around.

This estate with large plantations, in Louisiania, were accumulated by the owner, while in the business of slave-driving and negro trading. His name was Franklin. After his death his youthful widow married a gay leader in the fashonable world, known in the southern society of Memphis and New Orleans, as Joe Achlen. Under his direction the estate was improved and beautified at a cost of $1,000,000.

At the commencement of this war, it was had in contemplation by the Confederate officials, to purchase the estate and present it to his Excellency, Jeff. Davis; but they will probably defer making that munificent gift, until the Federal army is at a safer distance.

An intelligent chattel, who has been on the place twenty years, informs us that Achlen was a kind master. That when he visited his plantations in Louisiana, the negroes would welcome him at the wharf, and if it was the least muddy, would take him upon their shoulders and carry him to the house. But despite this fact, the negroes have somehow got the impression that freedom is preferable to slavery. So strongly are they impressed with the desire of owning themselves, that out of 900 who were on the estate and plantations at the commencement of the war, but five remain at the former place, and these with wages of $15.00 per month, while about the same number are at each of the plantations, these kept also by wages.

The death of Achlen occurred last fall; his widow is much of the time in New Orleans, but the property is neatly kept by what was formerly a part of itself.

One of those little incidents, by the by, which proves that truth is stranger than fiction, occurred to this negro who testified to the kindness of his master. When he was purchased for the estate he was separated from his wife, who was sold south. Neither knew the locality of the other, and nineteen long years passed by, when this war, which has made such an upheaval in the strata of American society, loosened the chains of the bondwoman, and true to the instincts of her nature, she started toward the north pole, to find freedom and her husband.

He says it was a joyful time when they met and recognized each other in the streets of Nashville; but we each have the privilege of entertaining our own ideas as to whether the race is capable of constancy and affection.

Even the Capitol has its mounted cannon, to protect it against the *citizens of Nashville.* During our stay in the city, we have had the pleasure of listening to a lecture by two Rev. Drs. of New York, and Brooklyn, in the Hall of Representatives, and *by moonlight.* They were to speak on the subject of emancipation and reconstruction, by invitation of Gov. Andrew Johnson, and Comptroller Fowler.

That afternoon, they had returned from the front, toilworn and weary, where they had witnessed the battle and ministered to the wounded of Resaca and Dalton. Upon proceeding to the Capitol, the moon was bathing all things without in her silver radiance, while within hid dark shadows, in strange contrast to an occasional silver shaft, through openings in the heavy damask curtains.

Queries revealed the fact that the Governor, Comptroller, and the man having charge of the gas fixtures, had gone to attend a railroad celebration, not having received word that the gentlemen had accepted the invitation to speak at that time and place.

Quite a number of gentlemen gathered in front of the speaker's desk, with some six ladies—the latter provided with seats; and after some consultation we found ourselves listening to interesting recitals of how "war's grim visage" had appeared to Rev. Drs. Thompson and Buddington of New York and Brooklyn.

And we could but think as we sat there in the moonlight, with most of the audience standing, what different audiences they had swayed at home, and how much depends upon time, place and circumstance in the life of a public speaker, and were glad to see that they could meet adverse circumstances with becoming serenity and humility. The novelty connected

with the scene, time and place, made it an evening long to be
remembered.

The Seminary building was used as hospital, then as barracks and since as soldiers' home.

The faculty of this institution, in their last advertisement
of its merits, previous to the arrival of the Union army,
assured their patrons that they would—

"So educate their daughters, as to fit them to become
wives of the Southern Chivalry and to *hate the detestable
Yankees!* "

The Medical College on Broad Street, is now a home and
hospital for the refugees ; and the filth, destitution, misery
and ignorance which exist among that class of poor whites
who have fled from starvation in Georgia, North and South
Carolina, Alabama or East Tennessee, must be witnessed to
be realized. We no longer wondered that the neat, industrious and comparatively well-informed negro servants and free
colored people of Nashville look upon them with the contempt
so well expressed by the words, "*poor white trash!* "

Brought up to think labor a disgrace, they will sooner sit
down in ignorance, poverty, and the filth which nourishes
vermin and loathsome diseases, than disgrace themselves by
work. Unaccustomed to habits of neatness and industry they
are singularly careless of each other's comfort, and neglectful
of their own sick.

The same week of our reaching this city, a family of refugees, nine in number, the parents and seven children, all
died, and of no particular disease. The scenes which they
had passed through, with the loss of home and each other.
with the native lack of energy which led them to succumb to
circumstances, rather than battle to overcome them, seemed
the only causes.

We will sketch a few of the scenes we saw in this home of

the refugees, prefacing, however, that some of the worst features we do not propose giving, either to offend ears polite or our own sense of propriety.

In company with the matron we enter the spacious building between two majestic statues, which stand like sentinels to guard the entrance, less efficient, however, than that "blue coat" who perambulates the walk with rifle and bayonet.

In the first room a gaunt and haggard face meets ours, with piercing eyes, from beneath an old slouched hood, and from a miserable bunk, whose possessor, within the next twenty-four hours, ceases to battle with consumption, and finds that "rest for the weary." She is now so restless she must be turned every few minutes, and stranger hands attend to her wishes.

" We were starved out," she says. " The Rebs tuk everything what they didn't destroy; and burnt the house."

" 'We,' who came with you? "

" Me two step-daughters. But they haven't been here these three days. I reckon they're tired o' takin keer o' me. It's mighty hard though to raise up girls to neglect ye when ye're on a death-bed."

What can we say to comfort her. Our heart grows faint when we think how incapable we are to minister to this one. Bereft of home, penniless, forsaken even by relatives, and in such agonizing unrest. Yes, but a happy thought comes now, if homeless, can she not better appreciate the worth of that "house not made with hands, eternal in the heavens"—if penniless, realize the enduring riches of the better land—husbandless and friendless, know better the worth of that "Friend above all others"—restless, the value of that "rest for the weary?" We tell her of all these, and she professes to gain new strength from our words to wait on the chariot wheels which so long delay their coming.

On another bunk is a wretched woman, who is drowning

7

sorrow as usual in the stupor induced by opium. We have now no message for her.

See that little chubby child, of perhaps three years, whose little flaxen head, has made a pillow of the hard hearthstone, and is soundly sleeping. That is a little waif—nobody owns it. It has neither father, mother, brother, sister or other relative in the wide world that any one knows about. Pity, but some one bereaved by this war would suffer this little one to creep into the heart and home and grow to fill the place made desolate!

Here is a tall, well-formed girl, of perhaps twenty, with a perfect wealth of soft, glossy, auburn hair, of which any city belle would be proud, but it is in wild disorder and just falling from her comb. Ask her, if you choose, what is that eruption with which her hands are covered, and which appears upon her face, and she will as unblushingly and drawlingly tell you, as though your query were a passing remark upon the weather.

Here are three other girls sitting upon a rough board bench—the eldest, a bright girl of about twelve, is making an apron for her sister. Do you wish to hear her story?—if so, listen.

"Me an' me mother an' me two sisters come from East Tennessee. The Union army come to our place first, an' they burned an' destroyed a great deal what they didn't take away, and after they left the Rebs come an' did the same, an' so between 'em both they left us all starvin' through the country. Then the Unioners come agin, and we followed 'em, an' they sent us here. While we were on the boat it was powerful open an' cold-like, an' me mother tuk cold. An' she looked like she was struck with death from the very first, an' the doctor told me I might just as well make up my mind to it, first as last, an' make her as comfortable as I could. So I tuk keer o' her, day an' night for two weeks, an' brought her every thing she wanted, oranges an' sich like, till she died. I thought when my father an' other relatives died that I tuk it powerful hard, but 'twas nothin' like losin' me mother. While she was sick me two little sisters had been livin' with a cousin o' mine ; but I hearn tell he was treaten 'em mighty bad, so I wrote a note to the captin an' told him I wanted to come here and see to the keer on 'em myself. An' he said I might, so I comed yesterday."

We leave this room for another. There a sick boy of fourteen is lying on a bed of rags, who is recovering from measles. Hear his history.

" We lived in East Tennessee, an' my father nigh onto the first o'the war, wanted to get to Kaintucky and jine the Yankees, but the Rebels tuk him off to Vicksburg and made him jine them. Then when the place surrendered to the Yanks, about half on 'em jined them, an' my father 'mong the rest,— jest what he'd been wantin' to, for a long time.

But they burned and starved us all out to home, an' we left thar an' come har whar we could git suthin' to eat. Me an' me mother an' me little brother what's only six year old come. But me mother was tuk sick an' died here three week ago. I hearn right after, that my father's regiment was ordered some whar else, an' I don't know whar he is. She knew what company an' regiment me father was in, but I was sick when he sent word about it, an' he don't know whar we air. Mother nor he could'nt write, so we've no letters nor nothin' to tell. May be he's dead, an' we'll never hear of it, or if he lives he'll never find us."

It is a sad case, but we comfort him with the hope of what perseverance and a little knowledge of writing may do for him, and pass to another.

Here is a young man, dressed and lying upon the outside

of his bed, whose foot and ancle are encased in a wooden box.
His temperament partakes largely of the nervous sanguine.
He has an open, frank, intelligent countenance, speaks rapid-
ly, and with a short, joyous, electrical laugh.

"I was raised in North Carolina," he says. "I was'nt a Union man at the first—nor a Confederate either, well about half
an' half, I reckon.   But we'se all obliged either to run away
from our families an' leave 'em to starve, or hide with 'em in
the mountains or jine the army.   So I concluded to jine ; an'
I've been in Braggs army mor'n two years."

"Why did you leave it," we asked.

"Well the fact was I begun to think *sure* we was in the
wrong, else we'd fared better'n we did.   For I've allays allowed the Lord would prosper the right ride.   So when I
found that I had to march or fight hard all day, an' have
nothin' more to eat for the hull twenty-four hours, than a
piece o'bread the bigness o'my hand, an' a piece o'meat only
as large as my two fingers—an' have been so hungry for
weeks that I could nearly *eat my own fingers off*, I concluded
to desert and try the other side.

My brother-in-law left Lee's army about the same time I
left Bragg's.   I was to meet him and my wife, at his house in
Athens ; but when I was coming on the train from Charles-
ton, I saw another train coming that ran into ours, and I
jumped off and broke my limb.   So I could'nt go there, and
they brought me on to this place.

I've enough to eat, and have good care, and should feel
right well contented till I get well, if I only could know
where my wife Martha is.   I've sent two letters, but I
can't hear a word.   I've got a letter written to my brother-in-
law about her now—its lying there."

And he points to a rough board, one end of which rests
upon his bunk, and the other upon an empty one near, and
which serves him in place of a stand.

"Its been waitin' a long time" he adds, for I hav'nt a postage stamp on it. We were just married when the war begun, an' we had a fine start for young folks, but I let my gold and silver go in gittin' settled, and the Confederate money's worth nothin' here, so I hav'nt a penny to use."

The letter was put in the office, and he was supplied with stationary and stamps during our stay. He wished more added to his letter and we wrote what he dictated.

" It's the *first time* I ever had anybody write for me," he said proudly. "I generally *do my own writin'*,—an' readin' too," and he glanced toward some books he had.

"An' you may be sure," he added as we left him, "if I get well, an' my wife Martha is lost, but I'll spend the *rest o' my life huntin' but I'll find her !*"

8

CHAPTER IV.

WEDNESDAY, April 13.

Entered upon my duties to-day, as lady nurse of two divisions of tents at Small Pox Hospital.

Not obliged to come here, but have accepted this most disagreeable place, as there are so few who are willing to take it. Expect to be quite confined to the place; and the hope of doing good in a position which otherwise would be vacant, is the inducement.

The Hospital is about a mile out from the city, and near Camp Cumberland. It consists of tents in the rear of a fine, large mansion which was deserted by its rebel owner. In these tents are about 800 patients—including convalescents, contrabands, soldiers and citizens. Everything seems done for their comfort which can well be, with the scarcity of help. Cleanliness and ventilation are duly attended to ; but the unsightly, swollen faces, blotched with eruption, or presenting an entire scab, and the offensive odor, require some strength of nerve in those who minister to their necessities. There are six physicians each in charge of a division. Those in which I am assigned to duty are in charge of Drs. R. & C. There is but one lady nurse here, aside from the wives of three surgeons,—each of whom, however, has her special duty.

Mrs. B., the nurse, went with me through the tents, introduced me to the patients and explained my duties.

April 14.

A woman and boy died in my division last night. The woman left a little child, eighteen months old, which is inconsolable. The father, a soldier, wishes to take the child away, but was not permitted to do so or to see it, for fear of contagion. It is to be kept to see if the child has the disease. [It did not, and had no scar from vaccination, such queer freaks the disease takes.]

The boy, an Alabamian, told me yesterday he was getting better. He had been sent here with measles, recovered from those, but the small pox did not break out. He died easy, and said he was "going to Heaven." I write his people to-day, via Fortress Monroe. His name was G. B. Allen, of Rockford, Cousa Co., Alabama. One man died yesterday, to whose people I have written to-day. Another died to-day. The mortality here is great. Said one patient to me:

"People die mighty easy here."

I asked in what way, he meant.

"Oh," he replied, "they'll be mighty peart-like, one minute, an' the next you know, they're dead!"

This is true, and I find so many who were sent here with measles, recover from those, and die of small pox. *Sixty cases* of *measles* were sent to this hospital in *one month*, as I learn from the lips of the surgeon in charge himself, Dr. F. These are sent by the several physicians of Nashville. The fact itself speaks volumes, but to stay here and see its effects day after day in the poor victims of such ignorance, impress one with a sense of the importance by the medical faculty of distinguishing between the two diseases.

SATURDAY, April 16.

I find many very interesting cases here, some of which shall wait to see the finale before making note of them.

What seems to me a strange feature, as I become more familiar with death-bed scenes, is the fact that so few know they are dying or are even dangerous, but persist with the last breath, or until the last struggle, that they are " getting better."

One poor young boy from Georgia, by the name of Ashman, who must die, although he eats nothing except a few canned peaches and milk, which I carry to him, will tell me sometimes when I go into the tent, that he is expecting a can of peaches every minute from home, and at another that he has just heard that his mother is in town, and that if he really knew she was, he would'nt lie there a great while before he'd be hunting her up. At another, he asked my name and State, and whether I took him to be a man or only a little boy. He is a slight little fellow of about 18, but in answer to the question I told him that of course I considered one really a man who could be a soldier and fight for our country, and who could be so good and patient while sick. To-day he called me to him, as soon as I entered the tent, and asked if I " could'nt discharge him to-day—that the doctor had told him to ask me about it, and that whatever I said he might do."

I told him that I would discharge him just as soon as that limb of his got well, and reminded him that he would want to be able to walk to the cars before starting home. He has a bad abscess on his limb, from which the doctor says the flesh is sloughing, and he does not expect him to live through to-night. And yet the boy wants me to "write to his mother in Atlanta, Georgia, and tell her to write to his aunt Shady, in Butler," that he " has been sick, but is getting better."

One man—G. W. Crane, of 3d Missouri Infantry, and who is called Major, was given up the day before yesterday by Dr. R.

He complained greatly of his throat, and I have since kept wet bandages on it, greatly to his relief. I asked permission of the doctor to do this, and advice as to telling him of his danger. He thought it would be well to do so, as he might wish

· to make some business arrangements. It was a most unwelcome task, but I believed it best; and first, asked him if he would like a letter written to his people.

"Oh no," was the reply, I shall be able to write myself in a few days."

"Perhaps you may," I said, "but we are all in more or less danger when sick." Adding as gently as possible, "How would you feel about it, if you thought you were not going to get well?"

The queries seemed cruel, but I knew he had loaned a gold watch and money to a man, and thought he might wish to attend to that and other matters. But he said decidedly "I do not think anything about it, as I have no doubt I shall soon be up again. And Madam," he added politely, "it would afford me great pleasure to talk with you, if I were feeling well and in good spirits you know, but my throat is so bad it hurts me to talk."

After this rebuff, and being really undecided as to duty in the matter, I left him. Yesterday I found him living, but evidently near his end, and I felt that I ought to let him know his condition. First, I asked as before about writing letters, when he said with great difficulty that he did'nt wish to talk with me as it distressed him to speak. I then said I would only ask him one or two questions and then leave him, and I said :—

If the doctor and all thought you could not live, would you wish to know it?"

He said "No," decidedly.

"Well then," I said "I will not trouble you any more, but if at any time you wish letters written, you can send me word by the nurse."

I left him and he died in about an hour. He called for water, but as the nurse raised him to give it, he exclaimed "I am dying," and then gave some incoherent charge, in which the nurse distinguished the words; "the lady" and "a letter."

His request has been complied with.

Mrs. F. was relating a similar incident to me the other evening. Dr. F. was at the depot in Nashville, when an old acquaintance was found there, who had been ill, had received a sick furlough, and was to take the cars for home. He was so feeble, he was persuaded to go to a hospital to remain over night, and take the train next day. In the course of the evening there was a change, and the physician knew he could live but a short time. He knew also that were he aware of the truth he would wish to send some message to his family. The man was speaking of his home and laying plans for the future, when the physician asked if he should'nt write a letter for him to his wife.

" Why no," he replied, " what need of that when I'm to start home tomorrow ?"

" You may not go then," said the doctor.

" Oh, yes," I must start tomorrow," was the reply.

The surgeon did not answer immediately, but was sadly thinking how to do so, and regarding the countenance of his friend, when the patient, who was about talking more of his plans, suddenly paused upon observing the expression of the surgeon's face, and earnestly asked :—

" Doctor — you do not think me very sick, do you ?"

" I do," was the sad reply.

" But doctor you don't think me dangerous ?"

" I think you a very sick man."

He lay silent for a few moments while thought was busy, and then asked :—

" *Am I about to 'cross the lines,' doctor ?*"

Tears, and the simple " I think you are," was the answer.

Then was business arranged, messages given, and they were alone again. Then he said :

" Why, doctor is this all that death is ? It's nothing at all to die."

And thus he " crossed the lines."

SUNDAY, April 17.

Attended service in dining hall. Chaplain S. officiated, and spoke very well. At the close I gave him the message sent by two sick men in my division to visit them. He promised to do so, but though he had to pass the tents where the men were, in going to his room, he did not do so. Am sorry, as the men may not live. He may have forgotten it, and if the men are living tomorrow, will remind him of the same. But I think it strange that he has not visited any one in my two divisions, when so many have died.

Three more have died since yesterday forenoon. Geo. W. Boughton, — Co., 2nd Batt. Vet. Res., Nelson Correll, of Co. B. 13 Tenn. Cav. and young Ashman mentioned in previous date.

One man, who is nearly given up by physicians, says he has been through the Mexican war. He is sergeant and will swear one minute and pray the next. He declares he always *has had* his own way, and *will have* it here. He is delirious part of the time, but like some others of that class thinks everybody crazy but himself. If it is his sovereign will and pleasure to get out of bed and walk about *en dishabille*, or take a trip over the mountains on some secret service, for which he fancies there is a war steed just outside,

" All saddled, all bridled all fit for a fight,"

he thinks the nurse is slightly out of his head to show so little respect to a superior officer as to threaten to tie him down to his bed. It has been necessary with him and others. He,

and another man who lay at a little distance, were both delirious last night, and had an argument with each other—or
what they supposed was one, though it seemed difficult for the
nurses to vouch for its connectedness. But it is certain that
a considerable number of oaths were used, and each assured
the other, in plain terms, that he didn't keep truth on his side.
The sergeant, after much gesticulation and violent language,
threatened the other with a personal chastisement if he wasn't
more *reasonable* in his *statements*. He was about stepping out
of bed to put the threat into practice, when the nurse produ-
ced a rope to tie him with, if he wasn't quiet; upon which he
concluded to defer the matter. When he wishes water, he
will sing out in a stentorian voice, for the

" Corporal of the Third Relief ! "

Monday April 18.

One man, this morning, while I was taking the name of one
who had just died, to write to his friends, told me that people
throughout the whole land, will bless me for what I am doing.
Wonder if I am doing good. I cannot help knowing that
some will hear from their friends who die here, who other-.
wise would not.

There is a singular case in Dr. C's. division. Upon entering the tent the first day after my arrival, with reading matter for distribution, I inquired of a young German if he could
read that language presenting a paper. He said " no," I then
offered one in the English language, asking the same question
He said he could read, but didn't wish the paper. The next
day I did not notice him particularly, as he was sitting up, but
the day following found him lying in bed, and that he would
not answer when spoken to. While feeding another man
with canned peaches who lay near, the nurse said :—

" You cannot make that man speak to you."

" What is the trouble," was asked.

" Well, it is this," was the reply. He says that day before yesterday, when you asked him if he could read English, he told you a falsehood, for he cannot read at all. He has been dreadfully distressed about it ever since, and says the Lord has appeared to him and told him not to eat a mouthful, nor speak to any one except once a day, to the surgeon and myself, until he has forgiven him for the sin. He will speak to no one, not even the other nurse who has charge a part of the time, and says, he will not, until he gets religion."

" What is his name?"

" Oswald."

" Wouldn't you like some of these nice canned peaches, Oswald?" we ask, dipping up some of the delicious fruit. He looked at us smiling but with tightly pressed lips.

" These are very nice—they'll do you good, and we want to make you well as soon as possible. Won't you have some, Oswald?"

No answer.

" Not going to speak to me? Why only think—here's a man trying to get religion and be a Christian and he won't speak to somebody else who is a Christian. I've professed to be one these many years, and you won't speak to me! Now, if you could only read the Bible, you'd know that it says " speak often to each other. You cannot read, can you?"

He shakes his head.

" Well, it's a pity, but don't you see that if the Bible says so, you ought to speak, and don't you see that Christian ministers have to talk to sinners to teach them to be good—and if ministers talk to sinners, shouldn't sinners talk to Christians—don't you see that?"

" Yes, yes, I do," he ejaculated, seizing my hand—"I will talk to you for you're a Christian."

9

We gave him some peaches and left him.

The next morning, however, nothing could induce him to speak. He has continued thus ever since—five days and has eaten nothing. He received a forcible cold bath this morning with the promise of its repetition if he does not speak and eat. [This was continued till he both spoke and ate. But he was believed to be a hopeless monomaniac, and after some weeks received his discharge and was sent home.] It is possible that this is mere pretence and his object the same as that of another soldier of whom we have heard, at Jefferson Barracks, Mo. This one used to go daily with a bent pin for a fishhook, and sit for hours upon a stump on the hillside, waiting quietly for the bite which never came, at least in the estimation of others. He was the butt of ridicule for the whole camp, who, while they pitied him on account of his supposed insanity, could but laugh at his perseverance in *fishing* upon *dry ground*. He received his discharge, when flourishing it in their faces, he informed them that it was " now his turn to laugh, as he had received what he had all along been *fishing for—viz: a discharge !*"

TUESDAY, April 19.

Another change. I am to leave this hospital to-day, as a Miss P. from Chicago, who had been engaged for the place, and expected some three weeks since, has just arrived. I have become really attached to the patients, and on some accounts dislike leaving. It seems that Miss O. and myself were intended for Chattanooga or other place farther toward the front, but in consequence of waiting for Miss O., the place was filled before our arrival. I fear there may not be any other place open for me. And when I can go in so many hospitals and see sick men suffering from neglect or want of more help, I shall think it very hard if I cannot do something. Two other ladies have been sent back, with the assurance that there was no opening for them.

I have just been through the tents and introduced Miss P. to the patients. Many are feeling sad, or appearing and expressing themselves so, that I am going to leave. Received many warm expressions of gratitude from many for the very little I have been able to do for them.

In going into one tent, found one of the nurses just recovering from an attack of lockjaw. When able to speak, he told me that it had "followed him, like an evil shadow, for ten long years."

Then followed an interesting recital of the cause, which was a gun-shot wound in the spine from the hand of a brother in an encounter with a grizzly bear in the rocky mountain. He himself ran away from home at the age of twelve, to follow his brother in a hunting expedition. After the brother had fired, the bear sprang toward him, and with one stroke of his paw laid the flesh from the bone from the forehead down one side of his face and arm to the elbow. The ball had only grazed the spine of the narrator, and seeing his brother in such danger, who called to him to fire, he did so and fortunately the shot was fatal to bruin. Their horses bore them to the nearest settlement, and the brother's life was saved.

This nurse I had always observed as quiet, efficient, faithful, and a favorite with the patients.

· The sergeant mentioned last under date of the 17th, overhearing me say that I was to leave to-day, and that I did not know where I should be stationed, advised me "not to be going round from one place to another, but to join a regiment, as I would be in less danger from guerillas."

Northern people, who think that all Government employees fatten on commissary stores, ought to see the table which is set at this hospital. It is exceedingly plain; and it sometimes requires more moral courage than all are very long, capable of exercising, to inhale the odor of oyster soup, custards, pies, and sweatmeats, which latter are sometimes prepared for those who are convalescing, but very rarely bless the palate of those who prepare them, or daily to deal out the jellys, blanc-mange and canned fruit without ever tasting. An instance of this kind has occurred here which not only increased our respect for the surgeon, but amused us not a little.

The usual rations, such as tough army beef, baker's bread and stale butter, with muddy coffee, served in brown mugs, has been the diet for so long a time that it has ceased to be very palatable. To the steward perhaps this was particularly so, and probably thinking that we had been sufficiently industrious and self-denying to merit a treat, and as five boxes of canned oysters had just arrived as a present from the Christian commission, he ordered enough cooked for dinner, in addition to the usual fare, to give all, from the surgeon in charge to the servants, a taste.

" It will take but five cans for us," said the wife of the surgeon-in-charge to me, " while for the patients a meal, it will require twenty cans."

So she, with the wife of doctor R., who jointly had charge of the diet kitchen, prepared the oysters, and at the usual hour, those, with the hungry expectants, appeared in the dining-room. The soup had been partially served up but no one· had time to taste it, when the surgeon-in-charge walked in and took a seat at the table. Probably the peculiar odour of the oysters and the ominous hush at the table warned him to be on the alert for something unusual.

Unusually demure, certainly, was the manner of the one table waiter, as he proceeded to the table, with another dish of the forbidden food.

The surgeon might well have exclaimed with Cæsar, " *Veni*, *vidi, vici*," for smoothing an instant smile from his features, with a forced sternness he demanded : —

" What have you there ?"

" Oysters," meekly responded the servant, who as well as the rest of us, more than suspected what might be coming.

" Take every one of those from the table," said he, " and don't let me see anything of this kind again. There are too many sick boys up at the tents, needing these things, for us to eat them !"

The oysters *were* taken from the table we are quite positive, and furthermore, that that was the last we ever saw of them.

It was, however, respectfully suggested to the surgeon by some one that he make it convenient to dine out at as early a day as possible, and acquaint his wife and the steward with the fact some time previous. He didn't promise, however, and the oysters have never since appeared to us.

10

CHAPTER V.

WEDNESDAY, April 20.

Back in town again. I've done something but havn't the least idea what, to displease somebody and havn't the least idea who. Perhaps some one of my friends here, will, after a day or so find the important secret too burdensome to keep alone, and will share it with me.

Just think what it is, Hallicarnassus, to go abroad and see the world—and feel it too, for that matter.

But in order to think as little as possible of that terrible crime of which I've been guilty, before finding out what it is, am going to hunt up enough work to keep my head and hands busy in the hospitals about town.

Glad to meet my travelling companion, Miss O., again. She has remained at this home of the Christian Commission, engaged in the preparation of delicacies, which are taken out to hospitals, or barracks, as needed.

This building, to which we came upon our arrival, is a spacious three story brick, at No. 14 Spruce Street. It was deserted by a rebel banker just before our forces entered Nashville, who took nothing south, except his gold and silver. A man from New York, whose conscience permitted him to take the oath of allegiance, removed and stored up against the return of his rebel friend, the silver and glass service, curtains, works of art, &c., but left much fine furniture, such as massive sofa bedsteads, marble-topped stands, tables, bureaux, a well-filled book-case, writing table and piano.

In Secretary Stanton's own handwriting, we saw permission

given to occupy this building till the close of the war, to Mrs.
H., of the Philadelphia Ladies' Aid Society, "together with
other ladies who might be associated with her, in any benevolent enterprise having for its object the relief of invalid
Union soldiers."

She is confident he meant benevolent gentlemen, also, so
one half of the house is given up to the Rev. E. P. Smith
and family, who make a home for the delegates of the Christian Commission.

Thus are many of the private as well as public buildings
reduced from their lofty position of serving southern chivalry,
to the vile misuse of northern mudsills. "Oh, Babylon how
art thou fallen !" must be the lamentation of the Nashvillians,
as they see the desecration of their beautiful edifices by northern vandals.

"Oh ! the citizens here would tear us to pieces very quick,"
said Mrs. Smith, the eve of our arrival, "were it not for the
'blue coats' about. Our dependence is in those and the
guns of Fort Negley."

Evening.

Visited the Refugee Home again, this P. M. Saw some
of those mentioned in a previous date. As I entered one
room, a woman was bustling about in a great passion, and
picking up a few personal rags, while ordering her son to get
up and they would find a place to stay where she shouldn't be
"set to do niggar's work !"

She was a healthy, strong woman, and had been repeatedly
requested to make her own and son's bed, and assist in sweep-
ing or cooking for the numerous inmates. Indeed, I think
she had received a gentle hint that it might be as well to see
that her son and herself had clean linen as often as once in
two or three weeks, and that the use of a comb occasionally

would not detract from their personal appearance. But she had her own peculiar ideas, obtained from living under the domination of a peculiar institution, and didn't fancy being dictated to in the delicate matter of her *personelle*.

Upon entering what is called the lecture-room we saw several families and parts of families, which had within two hours arrived on the trains from Alabama or Georgia.

I found that some of these snuff-dipping, clay-colored, greasy and uncombed ladies " from Alabam and Gorgee," are as expert marksmen as any of our northern exquisites, as they deposit the " terbaker " juice most beautifully into and around any knot-hole or crack in the floor, and while they are at the distance of several feet. Its wonderful how they do it—am afraid I should never be able to learn.

We approach one woman who is standing by a rough board bunk, upon and around which are several children overcome by the fatigue of travelling. She, unlike the generality, is neatly dressed in a clean dark calico and sunbonnet, and wears a cheerful and intelligent look. She informs us that these are all her children—six of them, that her husband is in the Union army, only a few miles out, that he had sent for her to come here, and she expects to see him in a few days. She cannot write, for she hasn't been to school a day in her life, and she says :—

" An' that thar's suthin' you people hev' up north, thet we don't. Poor folks thar, hev' a chance to give thar children some larnin'; but them as owns plantations down our way, don't give poor folks no chance. Larnin's only for rich folks. But my children shan't grow up to not know no more nor thar father nor thar mother, ef I kin' help it. Ef this war don't close so's to make it better for poor folks down har, we'll go north. Thar's a woman what kin' write," she adds with an admiring glance to the other side of the room, " an' she's writin' a letter for me to my husband."

We glance that way, and see a youngish woman, whose entire clothing evidently consists of one garment, a dress which is colored with some kind of bark. She sits in conscious superiority, scarcely deigning to notice us, as we approach, while she is carefully managing the writing with one eye, while her head is turned half way from it, so that the ashes or coal, from the long pipe between her lips, may not fall upon the paper. Her air and manner are evidently intended to be regal, for isn't she the woman " *what kin' write ?*"

At a little distance sat a hale, broad-shouldered, stalwart man, who looked as if he were able to do the work of half a dozen common men, who inquired of us, where " Hio was—if 'twas in Illinois "—and whether if he went to either of those places he would be " pressed into the service." In reply, we informed the gentleman that " Ohio was *not* in Illinois, but that if he went to either, he would probably have to stand his chance of being drafted, together with *other good loyalists*— with the physicians, lawyers, editors, and ministers. He did not reply to that, but his looks spoke eloquently,

> " For a lodge in some vast wilderness,—
> Some boundless contiguity of shade "
> Where war and draft come not.

Miss Ada M., the Matron of the Refugee Home, was, in our room this eve, and said that she was yesterday preparing some sewing for some young Misses, who were conversing earnestly about the Yankees. Finding their ideas rather erroneous with regard to that class of people, she made a remark to the effect that she was one herself.

" Why, you 'aint a Yankee?" exclaimed a Miss of fifteen dropping her work in blank astonishment.

" Yes, indeed, I am," was the reply.

" Why," said the girl, with remarkably large eyes, " I've

allays hearn tell that the *Yankees has horns, and one eye in the middle of their foreheads !*"

FRIDAY, 22.

Yesterday morning, Mr. F., a gentleman from my native State, Massachusetts, and who has charge of the Refugee Farm, asked if I would not like to ride out to the place,—they "wanted a teacher and perhaps I might be willing to engage as one, if not the ride and fresh air would do me good."

"Yes, I should enjoy it."

Then hour after hour passed away, with the fresh morning air, and not until at the dinner table did I meet my expected cavalier. He explained:

The fact was the poor old nag, which had been turned out some months before by government to die, like some other contrabands of war, wouldn't work—he was *free !* But he had confiscated another animal from Government and hoped he might not long say of that as in the nursery ballad, that

"The horse wouldn't go,"

as it was

"Time he and I were gone an hour and a half ago."

One, two and three o'clock came, and I overheard Lucy, one of the black girls, of about fourteen—though she doesn't know her age—laughing about "that thar Mr. F., who had been for two long hours, a curryin' an' pattin' an' feedin' that old horse with sugar, to coax it to be good: but I know by its actions it has never been harnessed 'fore a carriage in its life. For it acts, for all the world, like I did, when I ran away to find my freedom. I couldn't tell for my life, whether to go backwards or forward, to keep out of danger."

In answer to my questions, she tells me that she was "the

very first one that Lincoln set free in Winchester, but that as soon as she was gone, all the other nigs left."

Of course, her remarks about the horse were not very encouraging as regarded the safety or pleasure of the trip, even if he decided at last to go forward instead of backward. At half-past three, the equipage was announced in readiness, when, with a most self-denying spirit, I assured the gentleman, that I would willingly forego the pleasure, if the animal was not perfectly safe. But he was quite positive upon that subject, and as I perceived the appearance of the contraband did not indicate anything vicious or powerful enough to be very dangerous, we started. Had a ride of perhaps two miles upon the other side of the town, stopped a moment by the guard, then allowed to proceed a mile farther to the Refugee Farm.

This is best known to citizens as the Eweing farm. It was a splendid place, but has been nearly ruined by General Buel's army who camped upon it. Trees were felled, fences torn down, windows broken entirely out, and several fine outbuildings destroyed, such as a spring-house and conservatory, which I would like to have seen in its glory. Picked a beautiful bouquet of apple-japonica and pomegranate blossoms. Saw a " Butternut" planting cotton. He told me he expects, if the crop does well, to realize " one bale of picked cotton " from the two acres, which at present prices will bring $250. The yield, he said, was only about a half or a third what it would be three degrees farther south.

SUNDAY, 24.

Went out in an ambulance with Rev. Dr. D., Mr. E. and Mrs. H., Iowa State Agent, to hear the first named gentleman preach to a portion of the *fifth*, I think, Ohio Cavalry. They are camped on the Achlen estate. Saw a tree called the Red Bud and the mistletoe for the first time. The last grew on an elm. Secured specimens of each for pressing. Was indebted for the same to politeness of a gentleman who sported one bar.

Attended service also this morn and eve at Union Church; Rev. Mr. Allen officiated in the morning and Rev. Mr. Cramer this eve. The last is a young man and brother-in-law of General Grant.

MONDAY, 25.

The ambulance and driver were placed at my disposal this P.M., and I visited Hospital No. 1. I find changes here, but mostly for the better. Some have recovered sufficiently to be sent North. The "Alabamian," as he was called, who together with "William" was placed in my care, I am grieved to learn has "crossed the lines." He was getting better I was told, until one night he died suddenly of an ulcer on his lungs. William is dressed and walks around—is surely getting well, and talking of going home. Has had a letter written to his father and received a reply. Seems very grateful. The German suffered no more pain from the amputation, and is hopeful. The Norwegian has no gangrene in his arm now, and it is fast healing.

I find two or three new cases of interest. One is a middle-aged man who is suffering greatly from ulcers caused by scurvy. It is thought that he cannot live long; and he tells me that he isn't ready to die—that he has "been a bad man, that if the Lord will only spare him this time, he will live a different life." Another, a young man with fair skin, red cheeks and bright eyes, the victim of consumption, was moaning,

"Only to die at home with mother!"

THURSDAY, 28.

Am expecting soon to go to Huntsville, Alabama, as hos
pital nurse. Should have gone four days since, had not Gen
Sherman closed the way against everybody and everything
except soldiers, rations, gunpowder and pontoon bridges.
The road has been crowded with those for a week past. A
great battle is expected to come off very soon, some where at
the front. The Government has been pressing horses of
every description into the service to-day. The streets have
been crowded with teams marked "United States Transfer,"
those of " Q. M. D." and ammunition wagons.

This evening 600 horses have gone past our door, *en route*
for the front, where they are to act as scouts, I understand—
not the horses, though, I believe, but their riders.

General Sherman, himself, left for the front to-day noon.

During this time of waiting for a pass, rather than remain
idle, and also for the purpose of picking up some grains of
knowledge with regard to the " capacity " of the colored
race—which I believe a wealthy man said he would buy for
his daughter if she was'nt supplied with the article—I volunteered my services yesterday, as teacher in Mr. Brown's
school. This is held in the body of the colored peoples'
church, near the Chattanooga depot; Mr. B. is from Hamilton, Ohio, and is the pioneer here, in this work. There are
some 400 pupils and five teachers, all in one room. I supposed they were having recess when I entered, but found that
it was impossible to prevent them from studying aloud. It
seems it is practiced in the shcools of white children here, and
the great number in this one room, prevented such discipline
as otherwise would have been secured.

SATURDAY, 30.

The aptness of the pupils, as a whole, is really surprising.
11

Some have learned the alphabet, I am told, in three days, and others in a week.

It is said that all northern people who visit the school, very soon fall a victim to that fearful disease, known by the southern chivalry and northern copperheads, as "niggar on the brain." And I will confess my belief that were I to teach in this school very long, I might become so interested in some of my pupils I should sometimes forget that they were not of the same color as myself, and really believe that God did make of one blood all nations of the earth.

They present every shade of color from the blackest hue to a fairer skin than my own. It is often necessary to find out who the mother is before you know whether the person is white or black. The age varies from four to thirty.

The progress of some is really astonishing. One little black girl of seven years, and with wooly head, can read fluently in the Fourth Reader, and studies primary, geography, and arithmetic. who has been to school but one year. I inquired if any one taught her at home. or if she had not learned how to read before that time. " Oh. no. I learned my letters when I first came to school, and I live with my aunt Mary, and she can't read. She's no kin to me, and I havn't any kin, but I call her aunt."

Perhaps she never had any. or is related to Topsey, and if questioned farther, might say she " 'spects she grew." A boy of about twelve, who has been to school but nine months, and who learned his letters in that time, reads in the Third Reader and studies geography. Some are truly polite. The first day of my taking charge of one of the divisions, a delicate featured, brown-skinned little girl of about nine years came to me and said with the sweetest voice and manner :—

" Lady will you please tell me your name ?"

I did so, when she thanked me and said :—

"Miss P—— can you please hear our Third Reader this morning." It was not an idle question either, for the school is so large that now, while two of the teachers are absent, from illness, some of the classes are each day necessarily neglected. And so eager are the generality of the pupils to learn, that most of them are in two or three reading and spelling classes at the same time.

One might now not only exclaim with Gallileo, "the world *does* move," but add, and we *move with it*. For though but a little time since the negro dared not say "I think," lest the master might exclaim,—"*You think*, you black niggar—never you mind about that, I'll do your thinking for you," but would instead, say deferentially, with bent head and hand in his wooly hair, "Wall, massa, I'se been a studyin' about dat dar," is now learning to stand erect and confess that he *does think*, as well as learn to read and write.

One of the more advanced pupils told me that her father taught her to read and write before it was safe to let any one know that he did, or that he could himself read.

EVENING.

Eureka! That wonderful secret, like "murder," has "out."

I have been very cautiously, and little by little, and with many charges not to tell any body, informed of the terrible crime for which I was tried, convicted, sentenced and banished. while all the time in blissful ignorance of the crime itself. This is the way of managing affairs here, I am told, and it is called military style. I like it. It saves one all the trouble and worry of defending one's self. And that might make one nervous and excited. It saves also confusion in the mind of the adjudging party, the same as of a certain judge in Missouri, who having heard evidence on the side of the plaintiff, refused

to listen to that of the defendant, with the profound remark,
that "whenever he heard both sides he always got things so
mixed up, that he never could tell upon which side to give
judgement!"

But the grave charge, as ferreted out by some two or three
friends, of which I am accused, and to most of which I should
have plead "not guilty" had opportunity been given, runs
thus,—that upon a certain occasion, I presented myself before
the surgeon of the division and told him with an authoritative
air, that I wished he "would see that a certain patient had a
mustard poultice on his chest, for he wanted it."

This is my defence.   One morning, I found a man suffering
greatly with a pain in the chest from pneumonia, according to
the physician's diagnosis.   He was convalescing from vario-
loid and had taken cold.   He breathed very short, seemed in
extreme pain and begged for a mustard poultice.   I said I
dare not apply it without permission from the surgeon,
but would ask him immediately.   He was in another tent
—the third above, and while going there I recollected
hearing that some physicians were offended even by a request,
and hesitated.   Then thinking of the moans and apparent
danger of the sufferer, I proceeded.   These contradictory emo-
tions, I can now realize, gave an unusual brusqueness to my
manner, as I said :—

"Doctor there is a patient in the third tent below, on bed,
No. 9, who is in great pain and wants a mustard poultice.
Will you see if he needs it?   If so, I can make it."

There was a flash in his eyes, as he replied :—"*I* will attend to the man.   As for the mustard poultices, it is not
necessary that you should attend to them, as the men nurses
do that."

The patient did not have the poultice, but presume the physician gave him something which removed the pain, as it had

left him at noon. This trouble was caused simply by a mis-
understanding. He used the word *want* for *need*, so that
when I said the man "wanted" it—meaning he had *asked*
for it, he interpreted it so as to convey the idea of my assum-
ing the responsibility of saying, "he needed" it. He also
understood me to order him to "see" that the man had it,
when I simply asked if he would "see if he needed it."

I respect this physician and his wife, but wish he had been
certain of my meaning before reporting the speech to the
surgeon-in-charge.

There is also another little matter which I am certain had
something to do with my departure, but which it would
scarcely be policy for them to mention. It was this. The
next day after speaking to Chaplain S. about visiting those
sick men who had sent for him, and whom, though he was
obliged to pass the tents where they lay in going to his room,
he did not visit. I sent a slip of paper, saying in pencil, that as
he had probably forgotten it, and as they were anxious to see
him. I would remind him of this request. I received no re-
sponse to the same, although I am certain he received the
note, and the day passed without his visiting the sick men,
although, at noon, I saw him out for half an hour, engaged in
pitching quoits. I certainly did feel somewhat indignant,
when the next morning came, and I found from the lips of
the sick soldiers that he had not been in the tent; and I wondered, when I knew he had not been in to see a single sick or
dying soldier in my division since my stay, nor preached a
funeral sermon for the many who had died in my division
alone, what could occupy his time. I asked for information
of two of the ladies, and was told in excuse for him, that his
time was fully occupied in discharging the duties of clerk for
the surgeon-in-charge. So here was a chaplain neglecting
the sacred duties of his own profession, though amply paid

12

for the same, and earning more of the filthy lucre, to the neglect of dying men!

Thus endeth the defence. Mrs. Gala Days, you were entirely correct in your assertion that one must go abroad and see the world, to have " personal experiences."

SUNDAY, May 1.

This P. M., Miss O. and myself accompanied Rev. E. P. Smith to listen to his " colored preaching," as he termed it, in the same church in which is the school for the colored children.   It was a rare treat—and the first colored audience I ever saw.

Do not imagine a squalid, ragged, filthy audience ; but one where silks, ribbons, velvet, broadcloth, spotless linen and beavers predominated, with a sprinkling of beautifully carved or silver, and gold-headed canes, with about the usual proportion of fops to the canes that one may find in an audience of equal size, of our own color.   Some of these persons are free and own property.   But one would scarcely covet some of the ladies their silks and velvets, when she learns that it is purchased with the avails of extra labor at night after the day's work " for de missus is done."

But so it is.   And although the church was built some years ago with their money, yet it was held in trust by white people because " negroes cannot own property."

I have been repeatedly told that I would turn pro-slavery when I came south and saw how things really were.   I do not feel any of the first symptoms as yet, but quite the contrary.   Instead, I'm getting to believe that the day when the Emancipation Document was sent forth, was that of which it is said " a nation shall be born in a day," and I'm learning to think that this gospel, which is

" Writ in burnished rows of steel,"
and read by
" The watch-fires of an hundred circling camps,"
is the "word" which "makes men free," and will forever
strike the manacles from the oppressed bondsman.

One indignant white man, during the first prayer which
was made by a negro preacher, and in which he asked for
blessing upon the Union arms and freedom for slaves, left his
seat and walked the whole length of the church, with heavy
tread and with his hat on his head, while a voice called out,—
" Take your hat off!"

During the closing prayer the negro very properly prayed,
" Oh Lord, wilt dou give de people good manners and teach
'em right behaviour wen dey come into de house ob de Lord!"

The sermon was the Bible-story of the death of James and
the release of Peter from prison. It was told in a simple,
earnest, impressive manner, to a deeply attentive, impressible
audience. When he drew the picture of the angel entering
the prison, and taking Peter away as easily as though "his
chains were made of wax and a lighted candle was held beneath them, while the four quarternians—sixteen—soldiers
were powerless to act," one old man laughed outright, a
joyous, grateful laugh, others made their peculiar grunting
noise which no combination of sounds will give exactly,
while others shook hands and cried " Glory to God." During
the singing some women had the "power" so that they
passed round, embraced and shook hands.

Some joined the church, and the negro preacher told them
he " hoped that wouldn't be the last of it, and that they'd be
faithful and come to church ;" but that some joined whom he
" never could get a chance to set eyes on again, so that when
they died he never could tell WHICH PLACE THEY'D GONE
TO !"

I have forgotten to note in its proper place, that upon entering the church Miss O. and myself took seats in the only unoccupied pew in the body of the church. But Rev. Mr. S. beckoned us forward to a side seat by the pulpit. We took our seats there, but soon a neat, elderly negress came forward and said with a coaxing smile and voice, " Young ladies go up in de altar an' set—*you* doesn't want to set down here wid dese yere colored folks." We preferred remaining, and she urged the matter in vain. Soon an elderly mulatto man, probably a prominent member in the church, whose portly form was assisted in its waddles by a gold-headed cane, came forward and made the same request. But not being accustomed to the highest seat in the synagogue on account of our possessing a lighter color, we declined doing so until all the seats were filled and some must stand, when we did go ; but upon others coming in they also were induced to take a seat in the altar.

During the sermon Mr. S. related an interesting personal experience. He said that a year ago last July he was in front of Vicksburg, in that dreadful fever region—the Yazoo bluffs. He felt the fever coming upon him—he knew something of its workings—he was two thousand miles from anybody he knew. He said he " had been talking to the boys, to the sick ones in the hospital, telling them that it didn't matter where they died if they only had Jesus with them, and he found that on his back, and on his blanket, had come the time to take some of his own medicine." He said he " tried to do so, but found it rather hard to take. He tried to think that it was just as well to be sick there and to die and be buried on the Yazoo bluffs, and never see his family again ; but somehow he couldn't get in that frame of mind, but kept thinking he would much rather be at home. One morning, after he had burned and tossed with fever all night, Aunt

Nancy came and drew back the folds of the tent and said: "'Massa, how are you this morning—have you found the bright side?'

"'Well no, Aunt Nancy, I haven't found any bright side.'

"'Well, Massa, I'se sorry you can't, for Aunt Nancy never get in such trouble but she can find the bright side.'

"'Well, Nancy.' I said, 'I guess you've never had any very great trouble—guess you don't know what it is.'

"'Well,' said she with a sigh, 'may be I don't know what trouble is, but my old man was sold away from me down in old Virginny and I never see him any more, and then my son, the staff of my old age, was sold way down in de rice fields, an' I never see him any more. No—maybe I don't know what trouble is, but after that my last little boy an' girl was sold away from me, an' I never see them any more—an' now I'm getting so old I'll never go back to ole Virginny any more!'

"'Well, Aunt Nancy, that is trouble; but tell me how you managed to find the bright side.'

"'Well, Massa,' she said, 'when I see the storm coming, and the clouds are thick and get black and blacker, then I just *go 'round the other side of the cloud where Jesus is!*'

"'Then I turned over in my bed, with my face to the back of the tent, and said:

"'Come now fever, death and burial upon the Yazoo bluffs, if God wills, I am ready!'"

CHAPTER VI.

TUESDAY, May 3.

Spring has long delayed her coming here as well as northward. I could not be comfortable this P. M. in my room without a fire. Still, despite the cold, I have seen the blossom of a species of magnolia, which is very beautiful. It is in shape and size something like the African lily, and grew upon a tree the size of the common apple. It is of a peachblow hue upon the outside and white within, and with the mingled fragrance of the roses and lemon.

Aunt Nanny, the former housekeeper of the rebel banker who owned this residence, has just been giving me a highly interesting account of the scenes here when it became known that our forces were coming towards Nashville. It was on Sunday morning the news reached the white citizens, when they were on their way to church. And the streets were soon filled with half-crazed people flying here and there, women and children and even men running out of breath, and screaming, "The Yankees are coming," while the less excited ones were securing every possible conveyance to use for flight.

"We colored folks," said Aunt Nanny, "knew it in the night, and all de mornin' while de white ones was so quiet a putin' on dere finery for church, we knew it wouldn't last long. An' we was all so full wid de great joy, dat we'se a sayin' in our hearts all de time "*Bless de Lord*," "*Thank de good God*," for de "*day of jubilee has come!*"

"But we was mighty hush, an' put on just as long faces as

we could, an' was might' 'sprized when they told us of it. An'
missus she come runnin' back from the street wid' her bonnet
on her neck, an' the strings a flyin', an' she come to the kitch-
en and put up both arms, an' she said:—

"'Oh, Aunt Nanny, we'll all be killed! The Yankees are
coming! They'll hang or cut the throat of every niggar
that's left here!'

"An' after that she tried to have me go south with her, but
I told her I'd risk the Yankees a killin' us, an' I wouldn't
go."

Aunt Nanny is respected by all who know her. She is
neat, industrious, well-informed, although she cannot read, re-
spectful, polite, affectionate, *virtuous*, and a *Christian*. Her
husband is here, and she has one little daughter who is in my
division at school.

She tells me also that only last Sunday she saw the body
of a dead negro boy of about nine years who died from blows
received from his mistress. The cause was her anger that his
mother had run away in search of freedom. But the mother
heard of the illness of her child and returned in time to hear
him say that "the whipping his mistress had given him had
killed him," and to find upon his back the terrible gashes
from the whip, and bruises from blows.

I wish I had known of this before the child was buried.
Having the name of speaking my mind, it might be as well to
do so, occasionally.

Have just listened to a little incident which occurred some
months since. While Grant had charge of this department,
General Rousseau in his absence, issued an order to the effect
that slaveholders, who had taken the oath of allegiance, might
dispose of their slaves. One man from the country, accordingly drove in several slave women tied hand to hand. But
Grant had suddenly returned and countermanded the permit,

and he could not dispose of them. He got into his carriage and ordered his negro women to march home. They refused to obey. This was unprecedented insolence. He caught his horse-whip and was about laying it over their shoulders when the "blue coats" appeared as suddenly and as thick around him as if like fairies they had popped out of the ground.

"No whipping here! No whipping here!" they exclaimed, and the result was, he was forced to return alone, and they were slaves no longer.

WEDNESDAY, May 4.

A death in the house. Little Clark, the only son of Rev. E. P. Smith, aged three and a half years, died last night. It is a sad affliction. The disease, decline from measles. The funeral service was held in the parlor, this P. M., by Rev. Mr. Allen. His body was embalmed and is to be sent to the Sabbath School of the parish, over which Mr. S. presided, at Pepperill, Massachusetts.

To me it seems strangely touching—this trusting of the precious remains to the chances of travel, and so many miles away, to land in the throng of sad little faces to whose questioning glances he can perchance respond

"From the land o' the leal."

He was laid out in a child's military suit of light blue, with star-flowers, snow-drops, rose-buds, and leaves of the rose geranium. It was a sweetly sacred bequest to the Sabbath School.

SATURDAY, 7.

Have been to Hospital, No. 1, at the request of a mother whose young son had died there. She is in great anxiety to learn something of his last words, and whether he died a

Christian. I have just written her the facts, that there was no outward evidence of the same, but that she must trust the Good Father that it was " well with him."

All the patients whom I have mentioned in my journal, are better. Even the one with ulcers is improving. As for William he has applied for a furlough, and expects soon to go home.

Day before yesterday a girl came to school who had just the look and complexion of a snuff-dipping refugee. She, also, like them, wore a dress of the same color, derived from some kind of bark. Her manner was as listless and her expression as vacant. Wishing much to know whether she could claim our superior race as her own, or whether a few drops of the black blood in her veins had procured perhaps from her father and master the fiat—" only a nigger!" I made known my curiosity to one of the teachers, with my perplexity as to how I should obtain the coveted information, without wounding her feelings.

" Oh ! you need not fear that," was the reply, " they're used to it, and expect to be asked whether they're niggars or not."

I could not do it, however, without considerable circumlocution ; and commenced by asking if she could buy herself a book, whom she lived with, &c. After some time the questions eliminated the fact that though she didn't know whether she was free, or a " refugee," her own second name, or the age,—she did know that she had lived most of her life in Texas, where she had always worked out of doors, had hoed corn, and ploughed—that she lived with the same people now—that her father she had never heard anything of—that her mother was black, " though not real black," and finally that she herself was a " nigger,"—which nobody else could have told by her features or complexion.

A lady who stopped over night, on her way home from

13

Bridgeport, where she has been stationed with her husband in charge of sanitary stores, relates the following :

She said that sitting a few days since in the rooms where were the stores of the Christian Commission, she saw a woman, when half a mile distant, who had a long stick in her hand. She supposed that being weary perhaps, with a long walk, she had picked it up to serve the office of a cane. But after entering, and engaging for some time in jovial conversation and laughter with some neighbors, she found there, she made known her errand. She wanted to beg a shirt, pair of stockings and a coffin for her husband, and the stick was the *measure of his body.* My informant asked the age of the deceased husband, and she replied :

" Well, now, I never rightly axed him how old he was, but I reckon he mought be nigh on to thirty or forty ! "

TUESDAY, May 10.

My friend Miss O. is quite ill. We fear it may prove typhoid fever. Shall not enter the school again until she is better.

Last evening, had just seated myself to write to Mrs. Bickerdyke who had promised me a situation in Huntsville, Alabama, when she and Mrs. Jeremiah Porter, of Chicago, arrived on train from that place, bound for the front of Sherman's army, if they can procure passes.

To-day at dinner table, heard Rev. Dr. Thompson, of New York city, say that he saw 8,000 men march through the streets of Paris, at that farce entitled the election of Napoleon. I expressed my idea of the grand sight to Mrs. B. when she said that was not equal to what may be seen now. That there are " twice that number marching to the front now."

" Is there ? " was the surprised inquiry.

" Yes," she replied, " why don't you read the papers ? You

ignorant women in the army do ask such foolish questions!"

This is her style of speech, but she is a perfect hero in the army, among sick soldiers.

It seems that not long since she solicited and obtained contributions of fruit and vegetables for the soldiers who are suffering with scrofulous diseases for want of them. Some have arrived at this place, which she had ordered to be sent to Huntsville. She had left word while here some time since, with Colonel, or Captain Somebody, whose duty it was to attend to the matter, to forward them immediately upon their arrival. She also wrote the same from Huntsville, and still the fruits and vegetables came not, although she had learned of their arrival at this city, while the sick men in the hospital were suffering for the want of the vegetables, which were wasting from decay. This morning, she sallied out to ferret out the matter. In an hour or two she came capering into our room, where Mrs. P. was writing, and swinging her bonnet by the string, exclaimed:

"There, I've done it! I've said it! I've had it all my own way, for you wasn't there—addressing Mrs. P.—, to nudge my elbow and whisper 'be careful now, don't say too much,' or to tell the one I'm talking to 'Oh, she don't mean that.' Why what do you think I found?" she continued. I found those cars of vegetables moved on to a side track to *spoil for days*, while some of these officers have been sending on their *fine furniture to keep house with*, down to Huntsville. And after finding this out, I went to the office of this fine gentleman in shoulder-straps, and told him to send on those things in double quick-time, or I'd have his official head taken off. And I asked him if it needed a Miss Nancy to come and tell Miss Betty what her duty was, before she could do it."

FRIDAY, 13.

Miss O. since last date, has been daily growing worse. It seems she must have typhoid fever, and for one of her delicate health and sensitive nerves, we fear the worst. Her mind is at times wandering, and she dwells upon the scenes of filth and wretchedness she has seen among the refugees. At the commencement of her illness she was playfully told that she had "refugee on the brain." But it has since become too serious a matter to jest about, for she is sometimes certain they are in the bed with her; and this morning she told me of that "filthy refugee phlegm she spits up." She is a favorite in the house, and has every needed comfort and attention. I shall not leave for a hospital during her illness.

It was this subject which won from Mrs. P. the following : " A soldier at Fort Donaldson was wounded in the head, and was taken care of by Mrs. B. He was at times deranged, but got better and went to his home in Michigan. Afterward he became so bad as to require constant watching, and it was decided to take him to the insane asylum at Jackson. While on his way there, in the care of his brother, who was worn out with wakefulness, Mrs. B. entered the car. The insane man knew his old nurse, and she said the saying flashed through her mind that we should treat a crazy man as though everybody was crazy but himself. So she said to him :—

"Why you're taking your brother to Jackson, aren't you?"

" Yes," he promptly responded.

"How long has he been crazy?" she asked.

" Oh! he has *always* been crazy," he replied with emphasis.

So she told him she would help him watch his brother, and taking his arm walked back and forth in the car with him, and let the well man lie down and sleep. They had had much trouble with him, but he caused none the rest of the way.

She accompanied him to Jackson and has since heard that he is rapidly recovering.

SATURDAY, 14.

Mr. V., an acquaintance from Michigan, called on Wednesday. He is a secret messenger or spy for the Government. He wears the citizen's dress and a seven-shooter beneath the skirts of his coat, and has papers to show that he has a permit to wear the arms. His headquarters are here, and he goes on missions to and from the front. He says that on coming on the train from Chattanoooga last Monday the train was fired into. He saw one man in the act of firing and he returned the fire, and by the way he tumbled back into the bushes he had reason to think his own shot took effect.

He says the young lady at his boarding-house on Cedar Street, exhibited a pistol that morning, and said it was "intended to shoot a Yankee with; and that most ladies of Nashville carried one for the same purpose."

He told her in return that they might threaten, but that they seemed perfectly willing to accept a Yankee for a husband. It seems she has herself refused a wealthy citizen for a Yankee sutler.

Have found this eve that a Prof. P., a graduate of Yale, is about as nearly related to myself as the thirteenth cousin; and that he as well as myself can trace a relationship back to the "Mayflower." I think we each took a mutual dislike to the other from the first, and have been as coolly polite as possible; but this chance discovery will probably lead each to look with much leniency upon the faults of the other.

He tells me that with another of the delegates he has this P. M. been "the distinguished guest" of the 10th Tennessee Battery, which is stationed at the Capitol, and *very near* Governor Andy Johnson. That they were "sumptuously re-
14

galed with hard tack and *molasses,* and coffee with *sugar in it!*"

SUNDAY EVE, May 15.

Miss O. is very much worse. I did not sleep any last
night, and about three sent for Dr. F. She is suffering
greatly, and it is the opinion that before many days nature
may give up the contest. I cannot realize it, but fear I am
to lose this dear friend. Having had the exclusive care of
her and feeling quite worn, two ladies volunteered to take my
place to-day and eve. Sought sleep this forenoon but anxiety
prevented but little. This P. M. "Charley," as everybody
calls him, kindly prescribed fresh air and carriage exercise,
and we rode out about two miles to hear a delegate preach to
what is called "Anderson's corps."

It seems they enlisted with a promise that they should
constitute General Anderson's body corps but afterward were
forced into the common field service. They were mostly
graduates and professional men, and some have mutinied.
Wonder whose fault this was—this wrong done them? I
should be angry with them, even, should they tell me it was the
fault of the Government. The truth is that Red Tape,
which in its rightful province forms the firm ligatures which
keep in their proper places the different portions of the social
and military systems, is sometimes distorted from its original
use, and made to subserve the interest of petty underlings
and unprincipled officials. It is these who tell us that all
sin and high-handed wickedness which is wrought in high
places, must be "winked at."

"Red Tape is all right," said a poor boy in Hospital No. 8,
"if the commissioned officers did their duty, and had to come
under it the same as the privates. 'Tis the abuse of it which'
makes the trouble."

This poor boy had lain for seven long months in the hospital, while begging to go home, after his limb had been, as all the surgeons declared, *permanently bent nearly at right angles with his body.*

" Anderson's corps " is a fine and intellectual looking set of soldiers.

We also visited the 15th Colored Regiment and saw them on dress parade. Lieut. Col. —— accompanied me, and explained the changes and evolutions. He says that no regiment of our own color could so perfectly learn the evolutions, or a band learn to perform so well in so short a time.

THURSDAY EVE, 19.

My friend is somewhat better, but is very restless, and sleeps but little. She has been moved to a large, pleasant room in the third story where the air is purer, but the two large windows which open upon the front street and which must be open all the time to furnish air for the invalid, admit the continual tramp, tramp, tramp of the soldiers or horses, and the rumble of wheels through the long day and night. Her physician and friends think the only chance for her life is to obtain the pure air of the North and the quiet of home. It is in contemplation to send her as soon as it is considered safe for her to undertake the journey by water.

Have just returned from the Capitol, where I enjoyed the novelty of listening to the lecture in the Hall of Representatives, and by *moonlight*, which is described without date in Chapter Third. Rev. Drs. Thompson, of New York City, and Buddington, of Brooklyn, have just returned from the front to-day, and were witnesses of the battle near Dalton and Resaca.

Found a telegram from the brother of my patient upon my return. My last letter will reach him in reply.

CHAPTER VII.

On board the "Victory,"
Cumberland River, May 25.

So I am *en route* for Western Illinois with my sick friend. She was dressed for the first time since her illness to ride in a hack to the boat. Did not know of our going till about two hours previous. The hurry of preparation, and departure from friends, was trying to the invalid, and stimulants only kept her up to reach the boat. An excellent state-room had been procured and she was placed in the berth. We came on the boat last eve about five, the boat started about six and we are now steaming down the Cumberland. This is an excellent boat, there are scarcely any passengers, and everything for our comfort has been freely proffered, which, together with the gift of free transportation, evinces their sympathy with the Christian Commission and the cause of suffering humanity in general. Every one on the boat, from the colored servants and chambermaid to the captain, seem anxious to show every needed attention.

The invalid passed a miserable night without sleep, until after daylight, and is worse this morning. This writing is the product of seconds of leisure, between times of caring for her. She is full of sympathy for the sick soldiers, and the disappointment of having contributed so little for them, in her short stay, contributes not a little to her anxiety. She is one of those, who, if her physical strength kept pace with her ambition, would contribute largely to some labor of love peculiar to the philanthropist. As it is, she is one of our silent coral workers, and though her stay in the South has been short, yet there are those with whom her influence will go through life for good.

EVENING.

We passed Clarksville, Tenn., about nine this morning. Saw there the wreck of a boat which ran against the stone pillars one night, about two months since, causing the death, by drowning, of about forty Union soldiers. C. is a beautiful place. A fort guards the river entrance.

Later, passed fort Donaldson. Was surprised to find this a mere earthwork fortification, instead of some massive and strong stone or brick structure. It is four or five miles in extent, and on a high eminence overlooking the river. Our people now fear nothing from the river, but give more thought to the land approaches; with the enemy it was the reverse, and was the cause of our forces landing below the bend of the river, out of reach of the guns, and passing round and attacking them on the other side. The green-wooded hillsides were pointed out to me, on which are buried thousands of martyr soldiers, martyrs to the cause of our country, or that of ambition, or to false ideas of duty, but martyrs all the same.

An old lady of seventy, with crutches, came on board at Clarksville, who is going to Paduca, near which six of her children reside. She has lived near Clarksville, ever since quite a little girl, and has never moved over a mile in the time, till since the war has commenced, but thinks " its terrible moving times now." She thinks :—

" This was to be so, for Scriptur foretelled it, and she 'bleves *whichever is in the right* will conquer."

I inquired to what passage of Scripture she referred. " 'Twas that passage what telled 'bout ' wars and rumors o'· wars.' "

In the course of conversation, while answering her queries as to our destination, she informed us that she "*did* have a son an' a right smart lot of other folks, up in Illinois." In reply to the question of what part of the State they were in, she wasn't "sure now, 'bout that thar, but reckoned they might not be very far from Vandalia, or used to be, but now they'd mighty nigh all on 'em *died up* !"

THURSDAY, 22.

My patient slept well last night, and is better to-day. She told me this morning that she did not tell me how ill she was yesterday, but that she knew unless there was a change, she should never see home, and thought it would be so bad if she should die before getting there. I knew her danger yesterday, and know it still, but did not know that she realized it until she told me this. I scarcely fear death for her while on the route, as the excitement and stimulants will keep her up, but I fear she will have but a short time to stay at her earthly home before she goes to one better and more enduring, where is "rest for the weary."

EVENING.

Passed Cairo to-day ; and saw Fort Prentice. This also is merely an earthwork, or fortification, with one ugly looking. iron gun mounted and looking toward the widening of the river, like an open-mouthed watch-dog, ready to bay at intruders. There, the now swollen and muddy waters of the Ohio mingled with those of the Father of Waters, whose sandy color formed a striking contrast, and the line of meeting is plainly visible. It is, sometimes, the reverse of this, when the Mississippi is high, and the Ohio low. I there saw Bird's Point, and the residence of the man for whom it was named, while looking upon three States at once.

Passed Cape Geradeau about five o'clock. As the boat rounded to the shore, a coffin was brought down to the beach and then on board.

"There," said the Captain, "goes one more soldier home in a box."

But it proved to be the body of a Mrs. Bradley, who was drowned at the launching of a gunboat, at Carondelet, the 10th of last February. Other ladies were precipitated into the water at the time, but none drowned. Her body was found the 10th of this month, on a snag seven miles above Cape Geradeau. The body was past recognition, but there was a gold button with initials, which had been sent to her by her husband, and by which she was recognized. The finding of the body with mention of the button was made in the St Louis papers, which seeing, he came on, recognized the button, had the body unburied and with his little boy came on board with the body, which he is taking to his home in Cincinnati. He is a gunboat builder.

I saw two of those queer-looking Monitors to-day, at Carondelet.

FRIDAY EVE., 27.

The invalid is worse to-day. She suffers very much from exhaustion, but insisted upon being dressed before reaching St. Louis, as we expected to take the packet for Quincy to-night. But we arrived at the lower landing just as the packet was leaving the upper.

I dispatched a note to the agent of the C. C., as directed before leaving Nashville, and soon was in receipt of a note from his clerk, which was addressed to "E. J. P —— Esq." and commenced with "Mr. P., Dear Sir." Another note disclaiming the titles, and informing them that two lone women instead needed some attention, very soon brought Mr. Smyth on board the boat.

On Board the " Warsaw,"
Mississippi River, Saturday, May 28.

Everybody is very kind.   Mr. Smyth came on board again
this morning, and assisted in carrying my patient in an arm
chair to the carriage in waiting, and then accompanied us to
the upper dock and on board the " Warsaw," where he had previously secured a pleasant stateroom.

Transportation here is also freely proffered, and the captain gave the steward and chambermaid orders in my hearing,
to attend to every call for our comfort.   They are all so very
kind, and I am so thankful on Miss O's account.

Quincy, Ill., Sunday Evening, 29.

Last night we were called up in a hurry at twelve o'clock,
to change boats, as one of the engines of the " Warsaw," had
given out.   We exchanged for the " Northerner," a small
boat with inferior accommodations, and a slow sailer.   The
change was trying to the invalid.   The boat landed us at this
city about four, P. M.   A note was dispatched to the Rev.
Mr. King, Miss O's former pastor.   After some delay, Mrs.
K., with a gentleman, arrived in a carriage, and we were soon
at her comfortable home, and the weary invalid was soon
resting in a soft bed in a quiet room.   We have just had a re-
freshing cup of tea, and Mrs. S. sends me off to my room,
where I shall enjoy the luxury of a bath and a rest till morning, while she enacts the part of nurse.

Monday, 30.

Took express train, between four and five this morning, and
reached the home of Miss O. about seven.   The conductor
very kindly stopped the train so that the car in which she
was, came just in front of her father's gate.   There is no sta-

tion here, but the pale, wasted features of the sufferer were with him, as others, a passport to favor.

As the cars halted, her aged parents came out to receive their daughter as she was carried to the steps of the car. They supported her to the house, and not a word was exchanged between them and myself, except what concerned her comfort, until after she had been hastily undressed and placed in bed and restoratives administered. Then the aged doctor came forward and taking my hand, said tremulously :—

"And now is this Miss P——? "

"Yes, after so long a time," was the reply.

And then, with tears in his eyes, he pressed a kiss, *reverently* it seemed, upon my lips. The mother then kissed me also with tremulous tenderness. I wondered what induced them to welcome me in such a manner. I suppose they think I've been kind to their daughter, but if so, a good share of it sprang from selfishness, for I want her to live for my society.

MONDAY, June 6.

A week full of anxiety has passed on leaden wings. Have been ill myself, and necessitated to keep my bed much of the time, from the care, anxiety and wakefulness of the past three weeks. But that was a little matter, for I needed only rest to recover ; while for the invalid, we feared there was no balm in Gilead.

It was the second day after her arrival, that she came very near leaving this for the better land. She says she was so near, that she had blessed glimpses of that calm, serene, holy place. She cannot describe it. No beautiful " evergreen mountains of life," with their tops hid in the blue heavens, no gorgeous city with its spires and steeples pointing heavenward, no birds on rainbow pinions, or beautiful blessed isles of Beulah, sleeping on the broad waters in the hazy golden

15

distance, with mansions upon each, as though prepared for the souls of the blessed, which I have vainly pictured to her mind's vision, can be any comparison to that indescribable place, which gave such a feeling of "*holy calm and rest.*' Only a broad expanse of water stretching away beneath the azure of serene heavens, is the faintest emblem. She "saw the angels, just as plainly as she has ever seen me," who with their *balmy* wings, were

> "Round her bed and in her room,
> Waiting to waft her spirit home,"

as she often repeats to us.

She was then, and is now, sure that she knows what death is like, that its "strange coldness not like any other," came upon her, and she "felt the blood settling round her heart."

She had all day been complaining of a "suffocating feeling and pain through the lungs," but the feeling that death had come, came suddenly, and found many of her friends absent. A "pet brother" was away, and though willing to go, she wished "first to kiss darling brother good bye." And so she eagerly took spoonful after spoonful of wine to live until he might come. He did come and she wished to die in his arms, and so he held her until the moments grew into hours, and other friends were sent for, and more stimulants given to keep her for them, and still she waited upon the hither shore, or buffeted the swelling flood of Jordan. At one time she would be in an ecstacy of bliss with the beautiful vision, and at another, would feel that "Jesus had left her to tread the dark valley alone, and that thus she *could not* go." At one such dark moment she requested prayer, and a brother, himself feeble and sorrowing, offered up a broken prayer, and the light came to her gaze. Again it left, and the poor agency of my words was blessed to the restoration of her faith. Once more, only, was the light withdrawn. Then her brother was gone, and

she plead for some one to pray with her. No one was there
to do so, and she made the request of me. For years I had
not done that in public, but only in my heart and closet; and
had wondered whether if called to do so by some dying boy
in a hospital, I could discharge the duty. But the dying request was not to be refused, and taking her hands in my own,
while bending over her, I asked the dear Father not to allow
the *shadow* of death so to come between her and his blessed
presence, for though we were sure He was with her, we prayed Him to withdraw the darkness, which came through weak-
ness of the spirit, that she might, while crossing the dark
river, find Him her stay and staff, and might be permitted to
see his face, and know Him even as she was known of Him.

About midnight, she sank away to a quiet slumber, but
upon awaking in the morning *wept like a child*, that her clay
had not found the eternal sleep, and her spirit the endless
morning.

MONDAY, 14.

Another week has passed, but upon lighter wings, for the
wearing of alternate hope and despondency has, within the
last few days, been succeeded by the joyous conviction that
the crisis is past, and the invalid is slowly but surely convalescing.

Find my own health has suffered more than I had thought.
Shall not dare venture South again till the warm months of
the season are passed. In the mean time for medicine will
take a trip to southern Wisconsin, where, in the pleasant
homes of friends and whirl of happy meetings, health may be
found for a second trip to Dixie.

Have just heard two anecdotes, which I must jot down, before forgotten.

A *little* young soldier of this town, by name Breton, who

ran away from home and into the army, came home on a furlough, and staid a week over his time. On starting back, his
father took him by the hand and was about bidding him an
affectionate farewell, with a bit of parental advice, when he
cut short the matter by exclaiming :

"Good bye father, be a good boy and take good care of
yourself!" and he was gone. Upon reporting to his captain
for duty, the latter said :

"I believe you've overstaid your time, havn't you ?"

"Yes sir," was the prompt reply.

"What do you think I ought to do about it ?" said the captain. "Well I don't know, captain," was the reply, "unless you
put me on double rations!"

The second is told of a little neice of Miss O's. A brother
of hers has too little girls, the very opposite in character.
One is very amiable, quiet and gentle spoken, while the other
is a self-willed little spitfire. Both attended a "love feast."
Little angel took some of the bread and water, but spitfire
wouldn't.

"Why don't you take some of the bread and water, Adanine ?" whispered "gentle Annie."

"Taus, I aint a doin' to !" she said, spitefully.

When they reached home, their mother asked the little sin-
ner the same question.

"Taus, I didn't want to," was the vengeful reply.

"But why didn't you want to, Adanine," persisted the
mother.

"Taus, I *don't love everybody,*" was the confession.

"*Why* don't you love everybody, who is it that you don't
love," was the next query.

"Well," was the emphatic reply, "I don't *love de secesh nor
de debble !* "

At another time her father had to punish her, and he asked

" *what was* the reason she couldn't be a good girl, *why* she was so naughty."

" Taus—taus, I dit so tussin' mad," responded the little reprobate.

" Cussing," why Adanine, who learned you to say that word ?"

" Didn't *anybody*, I dess I know *some* tings don't *anybody* learn me, *'thout* its *de debble !* "

Think she will always be a firm believer in original sin.

UNITED STATES HOTEL, LOUISVILLE, KY.,
THURSDAY, Sept. 22, '64.

Last Friday noon, saw me in the city of Chicago, with trunk checked for Michigan. Entered rooms of North Western Sanitary Commission, and made myself known to Mrs. M. A. Livermore, to whom I had previously sent letter of introduction, from an old and mutual friend. Learned for the first time that a reply had been sent within twenty-four hours after its reception, with the offer of a position in a hospital at Rome, Georgia. This communication was rapidly given, and closed with the inquiry:

" Now, can you go: can you start on Monday ?"

Wednesday was preferred, and the result was that I returned to Harlem with the Rev. gentlemen who had accompanied me, where kind, though new found friends assisted in the preparation.

Yesterday, which was the day appointed, came to Chicago, and upon reporting myself in readiness for the trip, learned that General Sherman had issued an order forbidding any except soldiers going beyond Chattanooga; also that the hospital at Rome was soon to be broken up. However, as it was presumed a situation might be obtained in this city, Nash-

16

ville or Chattanooga, I was furnished with letters of introduction to dignitaries of the first two cities, and took the night train for this city, via Indianapolis.

A beautifully golden evening, and just cool enough for comfort. An excellent car and nice seat all to myself, luxuries appreciated all the more, as I may, before many days, be riding in a box car, through a country from the bushes or heights of which may whistle a bullet from an unseen foe.

Read two of the best letters in the world while watching the scenery and glorying in the triumphs of art over nature, as with the aid of a little fire and water, we sped swiftly over a corner of Lake Michigan, until the sun went to bed and the stars got up.

Then placing shawl upon valise, reclining in a very graceful position, and laying handkerchief over face so that my open countenance, if it chanced to be such, should not be subject to the vulgar gaze of Northern mudsills nor the lofty scorn of Southern chivalry, I sought the acquaintance of Morpheus. He was not easily persuaded, however, and between baby crying on one side and a political confab on the other, had only occasional glimpses of dreamland. The sun got up rather bright in the morning, but with a very red face, as if he were either ashamed of himself for sleeping so late or was out of all patience at the "goings on" down at the antipodes, or perhaps finding them so much better than ours grew red in the face at thoughts of coming back to us. Whichever it was could not determine; but was certain it was not from any sympathy with the copperheads. Well, the sun and I got up about the same time—myself a little in advance, and both just in season to get a view of the suburbs of Indianapolis.

The train arrived about six, would go at eight. Nearly two hours in which to hunt up an old friend. First inquired of omnibus drivers—no ticket agent to be found—then at nearest hotel for a certain Professor and Reverend. No one knew the residence of either. Landlords don't know, generally, when breakfast is ready and they can get seventy-five cents for a piece of tough beef, a cold potatoe and cup of muddy coffee. Called for a city directory, but only succeeded in finding the residence of the clergyman, which was a mile distant, and which might be a mile away from the friend I sought. So made the best of the matter, meekly ate an apology for a breakfast, meekly paid for it, and meekly requested a contraband of war to carry my valise and show me the train. Thanked the former piece of property, seated ourself, took writing materials and soon had a note written and dispatched to the Post Office, assuring friend C. that if she would enlighten the benighted understanding of the landlord, as to her whereabouts, she should be favored with a glimpse of the radiant face of a friend when next it passed that way, whether the time be a day, a week, a month or a year.

Travelled through clouds of dust until about three this P. M. before reaching this city.

FRIDAY EVE, 23.

According to instructions, immediately upon arrival, yesterday, proceeded to the office of Sanitary Commission. Pass into a hall. An open door upon either side. By one is posted this notice :—

"For sick or wounded soldiers, inquire here."
To which place we direct a man who had come from near Harrisburg, Pennsylvania, for a sick son who had been in Rome, Georgia. Turned to the other where several gentlemen were writing, and asked for Dr. Newbury.

"Do you wish to see the Doctor himself?" was the query.

" Yes, I have a letter of introduction."

" I will take it to him, and bring word whether he will see you."

In a moment he was down with the request to walk upstairs. Dr. N. met me quite cordially, offered a pretty army stool for a seat, and after finding out my purposes and desired destination, he said :

" It will be best to telegraph the surgeon at Rome, and he can, if he chooses, ask a permit from General Sherman for you to come there, or if the hospital is broken up, to go elsewhere with the patients. Be assured," he added, "that no effort shall be spared, and your interests promptly attended to. But as our orders are quite positive I prefer before you go farther that everything shall be properly done, so that we shall feel justified in calling upon Government for your support and expenses. In the meantime we shall see that you have pleasant quarters at a hotel, where you had best make yourself as comfortable and contented as possible until the matter is arranged."

He then called a young gentleman to whom he gave orders to accompany me to this hotel.

I was very grateful for this kind reception, but am certain that it was due to the influence of Mrs. Livermore, through her introductory letter. Had I come without it possibly my reception might have been similar to that of Mrs. Wittenmeyer last week, from General Sherman. His order had been issued, but she had pushed her way forward some way, and appeared at his headquarters at Atlanta. He saw her approaching and called out imperatively :

" Stop, madam—who are you—how did you get here?"

" I am Mrs. M., State Agent of Iowa and agent of Christian Commission."

" How *dare* you come," he angrily emphasized, " how *dare*

they *let you come*, after such positive orders as I have issued? *Go home*, madam—take the first train back, and don't stop this side of Chattanooga!"

She says she never walked faster in her life than she did to get out of his presence.

EVENING.

The young gentleman who accompanied and bespoke a pleasant room for me here, called this P. M. and accompanied me to Clay Hospital, "Branch C." It was M. J. Winder, "hospital visitor" of the Sanitary Commission.

The two Misses Wells are the worthy presiding geniuses of the hospital. Passed some three or four hours with the patients and found many interesting cases. Among which was that of a young soldier who, having been taken sick at some hospital farther south, had recovered sufficiently to start home on furlough, but upon arriving at this city, weak, weary and exhausted, he fell in the street and was taken to the Soldier's Home. Upon coming to, his mind wandered and memory was so weak he gave uncertain and contradictory statements with regard to himself, company and regiment. From this he was believed to be a deserter, and was put in the lock-up. Here Mr. Winder found him, when somewhat more rational, had him removed to the hospital, and had sent for his mother from Michigan, who has now had care of her son for some time. He is very talkative.

"They won't let me talk half what I want to," said the sick boy, "and I thank you for visiting with me; won't you come again in a day or two?" he inquired. And he added with emphasis: "Its done me a sight of good to visit with you—you're just the one to talk to us sick boys."

He was loud in his praises of the hospital visitor. He

is familiarly known as Henry. The look of his eyes, his nervous restlessness and the lack of sleep for several nights are unfavorable symptoms, still everybody seems to think he will recover.

While passing round and speaking to the patients, I found one man who was able to sit up but suffering from scrofula and heart disease ; and upon inquiring what State he was from, learned that he was from Cattarangas County, New York. And furthermore that he was a member of the 154th, which I had seen when it was starting for the South, and that his captain was an old school acquaintance—Captain Cheney. It was C. R. Brown, of Machias. While talking with him of other friends in the regiment, whose acquaintance I had made in school at Randolph, a young fellow approached and exclaimed :

" *You* from Randolph, New York ? "

And upon receiving a reply in the affirmative, with the addition of " more recently," he exclaimed, as he extended his hand with an emphatic nod,

" Why, that's *my home !* "

The action, manner and tone evinced the fact that he appreciated " home " as few beside soldiers can. And so it came to pass, that the rapid questions and answers revealed the fact that we had both been students at the same dear old Randolph Academy, and had each many of the same dear old friends. And I fancy we talked and felt as if we were the joint proprietors of all Randolph Academy—professors, teachers and students combined, and each was greatly rejoiced to meet the other partner in the concern.

While we were talking the matron came up and asked if he had " found somebody he knew."

" Why, yes," said he emphatically, " I've found an old friend."

"I thought you were looking better," she responded.

"Oh, yes," he replied, "I was nearly well before, and now this will cure me *sure!*"

He was quite right about our being old friends, as the trifling fact of our having never met before was not the least in the way. And our relationship was very much nearer than between those two individuals, who upon meeting in a foreign land, ascertained the pleasing fact that the dog of the grandfather of the one had once run across the garden of the grandmother of the other.

TUESDAY, 27.

The "Henry," mentioned under last date, was' suddenly called in the night to "cross the lines," but not into the country of an enemy.

In waiting for telegrams have passed some days at Uncle Sam's expense. Not a pleasing thought, but having a commission in my pocket authorizing me to take care of some of his sick boys, felt justified in so doing.

Yesterday met Miss C. A. Buckel, M. D., agent of Miss Dix. She also, per advice of the latter, had given me a call to come this way, which missive I had not received.

Rode over to the large new hospital at Jeffersonville, just across the river from Louisville, on the Indiana shore.

I can have a situation there, but with her advice, and my own inclination, shall visit Nashville.

Called this morning at office of Sanitary Commission, received permission of Dr. N. to go South, and a note from Mr. Thorne to Provost Marshal, who said a late order had requested the applicant for a pass to "apply in person." I went alone to headquarters and obtained the pass.

I leave the "City-of-the-Falls" for the "City-of-the-Rocks" to-morrow.

Colonel Ham, Indiana State Agent, whom I have met at the table of this excellent hotel, informs me that he received a letter last eve from the last-named city, stating that Forrest's and Dick Taylor's forces had combined and were marching upon Nashville. If that is true there will be wounded men to care for, and if a battle I want to be "there to see."

CHAPTER VIII.

HOME OF SANITARY COMMISSION,
NASHVILLE, TENN., Sept. 29, 1864.

The evening previous to my departure from Louisville I received a call from the hospital visitor and another gentleman, whom he introduced as Dr. Webster. I have since learned that he was sent out from Washington as chief of hospital inspectors, and is a brother of General Webster of Nashville, chief of Sherman's staff.

Dr. W. laughingly observed that he had called to offer his protection on the morrow, but presumed I was aware the offer now-a-days implied the desire to receive protection as well, when a trip to Nashville was anticipated.

The gentleman was informed that it would be a pleasure to bestow protection so far as a seat in the ladies' car was concerned, but that I should expect to be the recipient of the same should the train be attacked by guerillas.

Upon arriving at the depot yesterday morning all were ordered to take satchels, baskets, bandboxes, &c., forward to be examined with the trunks. But upon offering the keys of my valise and trunk to the inspector, he said:

"I guess you havn't any Government property you're taking south to sell," you're a member of the Sanitary Commission aren't you? looked at my pass, put a Government stamp on each article, and let me go without farther ceremony.

But all did not fare so well. The trunks of many were thoroughly searched; and I heard one lady, who came into the car just before starting, say:

17

" They *would* persist in diving into the very bottom of my trunk !"

Nothing special occurred on the route, and the time passed quite pleasantly in conversation with Dr. W. and a lady teacher whom I had met in the school at Nashville. Among other things the doctor spoke of seeing one Sabbath morning an aged colored man on some steps at Nashville, engaged in reading. He approached the student, and found him in possession of a Latin Testament. And upon inquiring if he could read it, the man humbly said he could not read much, having never had a teacher. But he did read and translate two or three verses quite readily. He informed the doctor that he had taught the colored people all he could for twenty years. That whenever one of his schools was discovered and broken up he commenced again in some other part of the city.

Arrived about seven last evening, and came immediately to home of Sanitary Commission, where doctor W. and wife are stopping. This is a pleasant place, on corner of Summer street and Capitol avenue.

Found Judge Root and lady, with their Sanitary family, at tea; and was not long in discovering that the table here is a place not only for the genial interchange of thought, and of jest and humor, the life of which is the judge himself, but also for the gathering of precious gems of knowledge, ranging from those of philosophical, geological and botanical science, to the latest news from the front, and the sayings and doings of our secesh neighbors.

As instance of the latter, we are informed that the widow of ex-president Polk has been informed that she can purchase coal of Government on the same condition as other citizens of Nashville,—by taking the oath of allegiance. That she has subsequently tried to get it at other places, but failed. That at one time she remarked that her "husband had been President of the whole United States, and that she cannot divide her sympathies and give them to any *one* party."

SATURDAY, October 1.

As result of introductory letter to Judge R. and his influence, had offer of situation in the diet kitchen of hospital No. 1. This was accompanied, however, with the proviso that I must be able to say to this one, " cook so much of this so long, and this so long," and also with the word, that he "allowed no lady under his charge to visit patients in the sick wards." Offer respectfully declined.

Yesterday morning called on Miss Annie Bell, matron of hospital, No. 8. She is very favorably known by all surgeons throughout the city ; and possesses a really noble and independent nature. She was at Gettysburg, and a winter at Harper's Ferry. She accompanied me to call upon dignitaries, and the result is a promise of a position for her cousin and myself in hospital No. 3, as soon as ladies' quarters can be fitted up, which are promised in about ten days. The surgeon is doctor Ludlow.

Yesterday P. M. had the pleasure of a ride about five miles out of the city, on the Gallatin Pike, to visit a field of cotton. We had the splendid team from head quarters, which consisted of four powerful black horses, and the only really fine-looking ambulance I ever saw. It had four seats, and the party consisted of five ladies, Doctor Webster and the driver. It was a delightful day, the air clear and balmy, and our steeds in fine spirits. We were obliged to cross on the rail-road bridge, the other being burned " to keep the federals back."

We passed the camp of the 13th " regulars," the last pickets, and drove down a road lined a part of the way with tulip trees, oaks, sycamores and magnolias.

We reached the cotton field, the driver sprang out, pulled away the fence, and the northern vandals were soon engaged in foraging each a handful of souvenirs. But the driver cautioned us through Doctor W. to make haste, and we found that he considered it quite a hazardous affair since leaving the pickets, three miles behind. There were country residences near by, and along the route, but they were violently secession in principle, and from the house, the owner of the cotton might easily have reached us with a bullet while we were engaged in the confiscation.

"He knows better than to do it, though," said the doctor. But Mrs. W. remarked that if her life was taken, it would matter very little with her afterward if he was punished for it, " It would not put her own head on her shoulders again and no other would fit them quite so well."

And as we were all more or less inclined to take that philosophical view of the matter, and considering also that our four splendid black horses might be a desirable item in the mind of some hardened bushwacker, we decided not to tarry long at the cotton, and the grass did not grow very long under the hoofs of our horses, until we were safe inside the pickets.

The cotton blossom more nearly resembles a white or cream-colored hollyhock than any other with which I am acquainted. It shuts at night, I am told, and does not re-open. There are small buds and large ones, blossoms in all stages, just formed bolls, and the ripe ones with the bursting cotton, all at the same time, and on the same shrub. The crop does not do so well in this latitude, this season, as usual. Indeed King Cotton rather disappointed me in his personal appearance, presenting rather a sickly and woe-begone look. That of his rival, wool, presents certainly a much more imposing aspect, particularly when the representative, like those frightful creatures at the north called Yankees, *has horns.*

TUESDAY, 4.

Helped Mrs. W. make two yellow flags, out of flannel, for the hospital train running between this city and Louisville, as they dare not run now without, for fear of being fired into. Two trains were stopped and burned near Fountain Head, this side of Bowling Green. They contained refugees who were robbed.

An order was received last Sunday, from General Sherman, to put this city in a "state of perfect defence." The probabilities of a battle here are a common topic of conversation. Should there be one, northern people are little concerned as to the final result.

Have made a visit to the Capitol in company with Mrs. Dr. W. At present, a New York regiment and six cannon are its protection. The lofty ceilings, spacious floors, broad flights of stairs and balustrades inside, and the whole exterior, with its gigantic columns, tower and graceful statuary are all of solid marble.

The senate chamber is less imposing, and the adornments fewer and much plainer than in the hall of representatives. In the end of that portico fronting the river is the vault of the architect, James Strickland, placed there as the tablet informs us, by an act of the legislature.

Visited the library and museum. The former seemed to me very large, but am told that it is not so considered. But Mrs. W. was occupied, I believe, in company with the wise wizard of the place, in consulting sundry yellow and ponderous

"Volumes of forgotten lore,"

to ascertain by what scientific name she might baptise a certain

18

shell and coral specimen she had picked up on Capitol hill, and which Doctor W. had declared "might have been turned over by the foot of Adam when walking with his children upon the beach, with the remark that he 'hadn't the slightest idea how old they were !' "

Being just now in a condition to sympathize with that young lady who had just finished at a fashionable boarding school, and who was surprised that she had "ever fagged through it all," and also that it was "astonishing that one head could contain it all," shall give the weak little head a rest from reading much about the world, till it has seen more of it.

There were several tattered flags in which I was much interested. One had been in the Mexican war, which was made and presented to the 1st Tennessee Inft. by the ladies of Nashville. Have been wondering how many of those same ladies now revile that flag, and prefer to know their loved ones are fighting under the banners of secessia.

I saw also such *beautiful* specimens of Tennessee marble, than which there is no finer in the world. A species of the red is used in the trimmings of the Capitol at Washington. I saw iron ore from the Ural Mountains, copper from North Carolina, tomahawks and axes made of stone, peace pipes and wampum taken from Indian graves, or their battle grounds. I saw a cast of Napoleon's head, the mummy of a man, and that of a sacred cat from Thebes, petrefied foam from the natural bridge of Virginia, a leaf from the Charter oak, an ambrotype of Samuel Houston, the original commission of General Israel Putnam, a spear from the farm of Osawatamie Brown, Continental money, the tooth of a mastadon, a horned toad, and a coat and hood of the skins of animals made and worn by Daniel Boone, of Kentucky.

What a very orderly and scientific inventory ! Think I

shall have to visit the Capitol once or twice more, and with paper and pencil, before I shall be at all satisfied.

SATURDAY, 8.

Visited the Penitentiary in company with a Miss I. H. Smith, from Quincy, Illinois, who has just been to Chattanooga with three tons of supplies designed for Atlanta. She had telegraphed General Sherman for permission to accompany them the rest of the way, but he replied in a kind note accepting those, but directing that they be placed in the care of an agent who would bring them safely. I learn by her that he has lately written one of his officers, in reply to a similar request, " Send always a barrel of pork, in place of a woman!"

The General is intensely complimentary.

Found it would be impossible to visit the military prison without a pass, with which we had neglected to provide ourselves. Were obliged to wait some little time for some one to accompany us, and in the meantime two ladies and a gentleman from the north, made a welcome addition to our party.

While waiting at the door, saw a party of about fifty Butternuts marched up close to the door, two by two, by a captain. They were halted and rations of bread and meat were dealt out, the first they had to eat in twenty-four hours. They were deserters, some from Forrest's forces. Saw a paper signed by two of them saying they were very anxious to be employed here by Government. They were marched away, and those wishing to go, will be sent north.

" We have in that yard about three hundred bushwackers and guerrillas," said the communicative guard.

" Ah, and what do you do with those ?"

" Well, we just stretch their necks for them a little," said he, with a self-satisfied smile, and with a motion of the hand

and neck as if in imagination he saw one in that very interesting situation.

"Just as you did Mosely the other day," we said.

"Yes, oh! he was a splendid looking fellow, fine features, well formed, black hair and whiskers, and straight as an indian!"

This Mosely was a guerrilla, who used to lay in wait by roadsides and kill the drivers of stray Government teams, burn the wagons, sell the horses or mules, and pocket the proceeds. He was hung a few days since.

There are now about one hundred and six in the Penitentiary proper, six or seven for life, and "the best men they have," and five or six are given the limit of the law short of that, which is twenty-one years.

We passed into the prison yard, the door was barred behind us, and we made the round of the workshops. First we entered the rooms where the native cedar was made into little fanciful pails and cups, in which the red cedar was dove-tailed into the white in wavy and curious patterns. I purchased one of these only about three inches in height. Various things for use such as pails, tubs, bureaus, tables, stands, large chests —nice for furs—and wardrobes are also manufactured from this beautiful red cedar.

It seems so strange to look at the men and to know that they must work on in *silence*, hour after hour, day after day, and year after year with a bar upon their lips. Of course to a woman it seems such a terrible punishment to keep one's tongue still. Isn't it horrible? I should think one's tongue would cleave to the roof of his mouth after a little.

Then we went into the tobacco factory and saw "the weed," from the time when the leaves are rolled and tied, to the pressing of the same, and the baking, to that when it is turned out "ter-bac-ker,"—a delicious cud for certain animals who are blessed with two feet, but which those with four never permit to pass their dainty lips.

" How is it about the health of those who work here all the time ? " was the query.

" Good," the overseer replied emphatically. " I was but sixteen when I first engaged in the business—was slender and weakly, but in a year's time was strong and well."

This does not prove, however, that he might not be just as well, if a carpenter or machinist, and his labor have been of some benefit to the world, instead of the reverse. Wanted to lower his self-respect a little by telling him so, but didn't.

We saw also the narrow cells where they sleep. One cell only was occupied, and by a maniac. He was chained by the foot, and standing in the open door with hands behind him. We were cautioned not to go within a certain distance. His position indicated that his hands were folded or carefully crossed, but we found afterwards that he held a club in his right hand. He watched us in silence with lowering eyebrows and hanging head, apparently measuring the distance between himself and us, with his small, black, malignant eye.

" Cannot I speak to him," inquired one of the ladies.

" Yes, you can, but I wouldn't advise you to," said our attendant. " You'd likely be sorry for it if you do. He never speaks to any one unless spoken to, but that easily angers him."

It seems that for years he was a captain on the Mississippi River, where he acted on the proverb that drowned men tell no tales with those whose purses he thought worth his care. He afterward became a highway robber on land. His term of fifteen years expired about a week since, and they have been trying to get him transferred to the Insane Asylum, but the officers of said institution object to receiving him on account of being made insane while here. He has been so dangerous that he has been chained constantly for four years. They dare not go near enough for him to get hold of one, and his food is pushed within his reach. Kindness they say, only makes him worse—treating those worst who show him favors.

MONDAY. 10.

Attended Union Church yesterday, in company with Dr. W. and wife. A very excellent and liberal discourse by Rev. Mr. Allen, from fourth verse of 3d Epistle of John,

" Walk in the light, in the light of God."

Called this morning on Mrs. James K. Polk to obtain some leaves and flowers for souvernirs of the place, to arrange on paper for a Sanitary Fair. Received very cordially by Mrs. P., who accompanied me to the grounds and cut the leaves and blossoms for me herself. She also presented a fine photograph of the place, taken from Vine Street, and showing the tomb of the ex-president.

Mrs. Polk has not entered society since the death of her husband. In person she is perhaps a trifle above the medium height, slender, with high forehead and delicate features, and bears marks of taste and refinement. Think she has passed through the ordeal of her former position with a true sense of its real worth in comparison with Christian duties and deeds of philanthropy.

WEDNESDAY, 12.

By this date I should have been established in Hospital No. 3, but just at the last moment, orders have come to the surgeon to prepare for the breaking up of the hospital as soon as possible. As the arrangements were not completed for our reception, it was thought best not to do so for only the probable space of three or four weeks. Miss Bell has accompanied me to other hospitals, but no immediate place offers itself, and I shall only wait until an answer to the telegram respecting a position at Jefferson Hospital is received.

Guerrillas murdered five negro soldiers night before last between this city and Louisville, near Gallatin, and set one thousand cords of wood on fire.

Last night three cars were burned near Bowling Green. Telegraph wires were cut. Previously there had been one thousand men sent to guard the road. Trains are almost daily fired into or run off the track.

This morning visited wards in Hospital No. 8 with the associate of Miss B. Some interesting cases. And while passing one bed was reminded of a conversation which occurred with the occupant when in this city last spring. He has now gone home. Upon inquiring his native State, after replying, he asked me the same question, and then said,

"Massachusetts—oh! that's an abolitionist State!"

"Yes," was the reply, "and I'm proud of the grand old hills, the free institutions and liberal sentiments of the Old Bay State."

"Well, I'm glad I don't hail from there," said the candid but smiling Buckeye.

"And I'm glad, if I was ever going to be laid up with this limb, that it happened before they sent niggers out to fight by the side of *me*. Didn't know this was going to be a *nigger war*, else they'd never got *me* into it!"

The hearer perceived he was in the gall of bitterness and the bonds of iniquity, told him so, and promised to call each day or two and devote an hour or so for his conversion. Did so subsequently, and found him always ready to converse pleasantly, but not a willing disciple. Am still deeply concerned for his future salvation.

Yesterday morning Dr. Woodward and wife left for home.

He was surgeon of 22d Illinois Volunteers. They have been stopping here since my stay. One evening last week, while the inmates of the Sanitary Home were seated around the genial fire in the parlor, the conversation turned upon the magnanimity of the soldiers, which it seems is not confined to our Union boys. It turned into another channel afterward, and some incidents were related, not in exemplification of the magnanimous, but very interesting, nevertheless.

The surgeon related that while near the battle-field of *Perrysville*—I think—one rainy afternoon, his son came to him with the word that two wounded soldiers were back of the hospital, near a swamp, who needed care, and whom, unassisted, he could not get away. Dr. W. went there with stretcher and attendants and found that one of our men had a shoulder shattered, and his companion, who was a Rebel, had a thigh in the same condition. The Union boy professed himself able to walk to the hospital, "but," said he, " I wasn't going to leave *him*, for I knew if I didn't stay and see that he was taken care of, he'd die to-night." He had somehow managed to take off his own coat and spread it over the other. The Rebel was put on a litter and carried, while the other, after having his arm put in a sling, walked to the hospital. Both had a limb taken off, and both died next day.

He said also, that while going round in the evening to ascertain who were in most need of help, and who could wait till morning, he came to one man whose arm was nearly shot off. It was a Confederate. The doctor had scarcely commenced the examination when the wounded man said :

"Doctor, I can wait, but I wish you would see what you can do for that man who was brought in with me—he is worse than I am and needs you more than I do."

" Which one is it," inquired the surgeon.

" Oh! it's one of your men—he lies there," he said.

"I'll take care of you first, I guess," was the reply.

"No," he persisted, "if you'll just put a string round my arm, so I can hold it better, it'll do well enough till after you take care of that man,—he's pretty bad."

"Well," said the doctor, "I'll take care of him first if you wish me to, but guess I'll give you a dose of morphine so you can sleep to-night, and in the morning your arm will have to come off."

"Well," said the noble fellow, "you needn't do *anything* for me till you've taken care of *him*."

Dr. W. did as requested; and both recovered.

The same physician told of one of our men who used to lie in his cot and read aloud from the Bible. One day he was passing the bed of one of the "Johnnies," when the latter said,

"Doctor, what book is that thar Yank readin' out of?"

"It's the Bible," said the surgeon.

"Well, I don't know nothin' 'bout readin' myself, but if you've no objection, doctor, I'd like to lie over thar nex' to him."

"Well," was the reply, "if the other boys are willing, I'll let you go there."

No objection was made, he was moved and used to lie hour after hour with his face turned towards the reader listening and asking explanations ; and after about two weeks he died.

Dr. W. also related a little incident which occurred on a march. They were passing by a farm-house, when the woman came out as General Paine was riding slowly by, and she called out in a querulous tone, "General—General!"

"Well, what's wanted," inquired the General.

"General, I want you should put a guard round my well—your soldiers are going to drink it all dry, so I shan't have any water for my family."

19

The soldiers were heated and thirsty with the long march through the dust and broiling sun.

" Won't you put a guard round it, General ?" persisted the woman.

" Yes, I will," said the General emphatically.

" Orderly—here ! "

That officer came forward.

" Orderly, put a guard round this woman's well, and don't you allow man, woman or child to come near it *till every soldier has had all the water he wants.*"

The same officer says that poor people often complain, and justly, that while a guard is set round the fine house and grounds of a rich neighbor, their own are over-run and pillaged, illustrating the passage that " to him that hath shall be given, and from him that hath not, shall be taken that which he hath." He says he knew of a place where three union soldiers were sent to guard a house, who were never seen or heard of afterward.

At Jackson a squad of soldiers were ordered to guard the residence of one, who, those soldiers were positive, was a rank secessionist. The house was burned down in the night, and the captain of the guard being questioned about the matter said he " guessed the lightning must have struck it." The house, *strange to tell!* was burned to the *negroes quarters,* which were saved. It is perhaps needless to add, that if the lightning did strike it, there was no thunder shower to accompany it.

During the conversation young Eddy Jones related the following as occurring on the train at Louisville :

The cars were about starting, when an officer came round to inspect the boxes, satchels and valises. Upon coming to one man who was sitting just back of the narrator, he found that he had a pair of pistols in the bottom of an old-faded carpet sack. This man was dressed in scant and short pants, old-fashioned coat, and steeple-crowned straw hat, and looking otherwise like a green country boy. Thinking to have some sport with the "greeny" he called out, sternly: "guard come here and put some irons on this man." "Here, hold on," said greeny, deliberately, and he took some papers from his pocket, which informed the official that he was ordering irons for the disguised colonel of the 58th Illinois.

Dr. W. saw also a man at Louisville, who was ordered to hand over the key of his satchel to the baggage inspector.

"There is nothing in the satchel except wearing apparel," persisted the owner emphatically.

" I must open it," said the officer, "its altogether too heavy !"

Upon doing this, were found, carefully done up in wearing apparel, five or six revolvers and as many boxes of ammunition, together with $300. The guard was called, he was marched off to the military prison, while he was informed that his property was confiscated, including the money.

Professor Hosford, of Hudson, Ohio, was present, and related the following, after the conversation had turned upon the condition of the freedmen. He had a conversation with a negro at Chattanooga, who told him of his liberation from slavery. Said he :

" Before the Yankees come here, missus used to tell us about other niggers leaving their masters, and axed what we thought of it, and we told her that *we'd* never leave missus, oh ! no, we thought too much of missus to do dat. But when de Federals was a coming into de place, missus got some baskets, and packages, and said we must carry um, an' we'd all leave. But we 'fused to go, an' missus, she had to go 'lone."

" Ah," said the Professor to him, " what did you tell her that you never would leave her, if you meant to, all the time ?"

" I 'tink it was right," replied the negro, emphatically.

"An' I can prove it 'cordin' ter scripter. For dough I can't read, I've hearn 'em read dis:" ' Agree wid *dine adwersary quickly while dou art in de way wid him, les' he takes you to de officers,* and *dey cas' you into prison !* "

Another negro at the same place told the Professor that he "Allays prayed an' prayed for de time to come when de colored people could worship God under dere own vine and fig-tree, when dey could stay in prayer-meetin' after nine o'clock at night, if deys a mind to, wid none to molest nor make 'em afraid. An' I'se allays▸believed de time would come, dough afterward I gets most discouraged wid de waitin', an' I never see any signs of my vine an' fig-tree a comin' till I seed Hookers' men a comin' ober de top o'Lookout Mountain !"

He had about the same idea of the working of God through direct agencies as a gunner of whom we have heard. He was behind his gun while the shells were bursting around him, when the chaplain approached and asked if he felt that Providence was supporting him.

"No," he replied, "I am supported by the 29th "New Jersey !"

FRIDAY, 14.

The negroes had a dance down stairs last night. I wrote several invitations for Miss Lu and Narcissus to Mr. so and so, dictated in this style, with variations:

"Miss Lu, wishes the pleasure of Mr. Baker's company round here, this evening, to a dance. Please come early,
          Miss Lu Palmer.'

All went down stairs for a little time, to see the performance. Eddy J. proposed that I ask the "musicianers," as aunt Polly calls them, for a " plantation break-down." He was commissioned to make the request for me, but the white-gloved and perfumed exquisite, assured us that he

" Wouldn't 'tink of such a 'ting, *heah*," and he gave me

such a commiserating and be-*nig*-nant glance and smile, as much as to say : " You poor white child, how I pity you for not knowin' what is expected ob dis 'spectable company ob colored pussons."

Before leaving this place must jot down something of my two contraband pupils.

" Well, Peter, what are you going to give us for breakfast," queried Judge R——, quizzically, of the little negro who waited at table the first evening of my arrival.

It was in a lull of the conversation, and just before the company rose from the table, so all eyes were turned naturally towards the boy, who bore the attack bravely and returned the compliment in full from his own, large, black orbs. He was well used to the quizzing from the merry-hearted Judge, and the pleased expression of his eyes and the exhibition of a double-row of the whitest ivory attested both, as he murmured, " I don't know."

Whereupon the Judge proceeded to name over a most bounteous and unheard of list of edibles, ending with that of " baked white fish." " And *be sure* Peter, that you remove *every scale* and pick *every bone* out of the fish before it is baked."

" Yes, sir," responded Peter.

" You'll attend to it, will you Peter? pick every bone out, before it is baked," said he, in a tone in which perfect authority and confidence blended.

" Yes, sir," lowly and submissively replied Peter, but with a merry twinkle of his eyes.

The next occasion in which Peter was brought particularly to my notice, was a day or two after, when, as I was passing along the lower hall, I came upon him and the other waiter at table. a girl of fourteen, named Narcissus, both of whom were trying to spell out the reading on a bottle of pepper sauce.

20

" Can you read ?" was the query.

" No, leastwise only a little, *wish I could*," added Narcissus, heartily.   And Peter said " I can read some, but I don't have nobody now, what'll listen to my readin.' "

" Do you have time to read ?"

" Yes, we'se a mighty heap o'time evenings, after the dishes is done."

" I don't know that I shall be here over three days," I said, " but possibly a week or so, and while here, will hear you read each evening.   What time shall I come to the dining-room ? "

" Right after tea," they said, " right after the folks has gone up stairs."

So that evening, I heard them read, unthinkingly, before the dishes were washed.   But as I was leaving the room, " Miss Lu Palmer," the elder sister of Narcissus, reminded me of the better way, by saying :

" Miss P——, if I were you I wouldn't hear these yere lessons, till they'd washed up their dishes.   They'd hurry if they know'd they'd have to wait till afterward, and you know dese yere colored folks don't like to work none too well, no how."

I confessed that she was right about the work being done first ; and thereafter it was dispatched with a will each evening, the " sooner to get at the lessons," as they said.

They have manifested the same spirit ever since, and learn rapidly.   Narcissus said one evening, " I don't know Miss — that I've got my lesson, but its all the time I could get, I've been a learnin' of it ; and last night after you'd done gone hearing us read, I studied the lesson right smart, and then dreamed about it all night."

And one evening while Peter was battling like a hero with the, to him, formidable task of spelling the word " occupy,"

I could but help wishing some of those believers in the universal stupidity and carelessness of the race, might have been listeners. Something very like the following they might have heard!

The word " occupy," is pronounced.

Peter hesitates a little, then with the voice and look of one who determined to make one good, bold attempt with his best judgement, and trust to luck, says O-k oc q-u-i cu p-i py, occupy."

" Oh, no ; don't you remember what other letter I told you, has the sound of k ? "

Peter *don't* remember, so *many* new things he has been told. " Well, now, think how you can spell " oc," without saying o-k."

No enlightenment dawns on Peter's mind, though he makes two or three bold attempts, to show his good will.

" Well, Peter, spell cow."

He does so correctly.

" Now don't you see that we could spell that in this way, k-o-w ? "

Peter sees that, as the word is spelled both ways for him and pronounced the same. It is then applied to the syllable " oc," and so much is accomplished. Peter has acquired a mastery, has obtained a new idea, and he takes a long breath for he has scarcely breathed during the time, and his eyes look as if proudly conscious that he had mounted another round on the ladder of knowledge. The word is again pronounced. The first syllable is spelled correctly, but " q-u-i p-i," is the only reasonable way the sinking Peter thinks the word can bo finished.

Teacher pronounced the syllable " cu," plainly, saying : " q-u-i spells qui, not cu, spell *cu*."

" K-u cu," responds hopeful.

" K you say, what other letter has the sound of k. ? "

He " don't know."

" How did you just spell oc ? "

" O-k ? " in a questioning tone, says the pupil.

" Oh, no, you've forgotten, you've so much to learn :" and the explanation is gone over again, and the sound of c for k requested for the second syllable, also.

He looks out of all patience at his own dulness, but heroically returns to the charge. This time he gets both syllables right, but ends as before with p-i.

" Now, Peter, you've worked like a major, and its all right except *one little letter i.* Now put on your thinking cap, and hunt up some letter to use instead of *i.*"

Peter " reckons a," will answer.

He is advised to spell lady, and does so correctly. " Now could not you have spelled that l-a-d-i and pronounced it lady ? "

He " reckons so," and is advised to use the same substitute in the other word. It is at last accomplished ; and after sundry mistakes in each syllable, during which he exclaims, ' Now don't tell me, I *will* have that," I'll get it right this time *shore*," the whole word is spelled correctly, and re-spelled repeatedly during the evening, and he enthusiastically exclaims :

" Well, that's the toughest word *shore* I'se ever a holt on, an' I'll never forget it long 's *I* live."

Peter's history is not uninteresting. Here it is : " My master's name was Jim Brazier, an' I lived eight miles from Tullahoma. My mother was sickly a long time, and missus wouldn't let her stop workin' no how. An' one day wen' she's so weak, she let a big pitcher fall out' de floor and brokt it, and master sent her to de whippin'-house, an' she died that night. I slept wid' her, an' she told me wen she comed to bed, dat she t'ought if she went to sleep she'd never wake.

An' in de mornin' wen I waked, she was stone dead. Dey neber said anyting to me 'bout what killed her, dey knowed berry well dat I knowed de reason. Atter de war brokt out, dey telled me dat I mustn't go near the Yankees, for dat dey " *had horns*." jist as if I'd not *sense* 'nough to know better nor *dat !* An' dey tole me I must keep 'way from dem, else dey'd cut off my ears and hang me on a tree. But arter dey'd whipped me and hung me up by my thumbs. for bitin' missus, when she had me down on de floor an' was poundin me 'cause I didn't sweep clean, I runned away."

" I'd been wid master three times wen he'd been to camp to sell apples and things to the Yankees, an' so I knowed whar to go. So one night I tuk one o' marster's hosses an' put a bridle on him, an' rode him most to camp, so near, I I could hear de pickets; den I fixed up de bridle, arter I got off, an' set him off on a right smart trot toward home, an' hid in de bushes. Den I waited till mornin', which comed pretty soon, an' I tole de picket I wanted to come in camp. He let me in, an' I'se roun' two or three days, wen Dr. Woodward said he'd see to the keer o' me, an' he has ever since. He brought me here. He's allays been right good to me, an' never gin me a cross word."

I found, upon conversing with Dr. W., that this was a truthful account, as far as could be ascertained. One morning, soon after, Dr. W. announced to Peter that his former master had just been hanged as a guerrila. The account was in the morning paper.

" Glad of it," said Peter, emphatically ; " I'd a ben glad ef dat ar' had a happened afore. He made me carry letters to the rebels tellin' 'em all 'bout whar de Yankees was, an' a pretendin' all de time to be a good Unioner. Hanging *good* 'nough for him."

This last, I also learned, from the doctor was the truth, for

Peter had guided our people to the hiding place of these clandestine letters, which were captured.

SUNDAY, 16.

Yesterday called at Rail-road hospital, also at hospital No. 8. At the latter place found one young man from East Tennessee whose father was shot when Lincoln was elected, and his house burned. One brother was killed at Gettysburg, and of the rest of the family eleven in number, a mother and brothers, he can obtain no trace. He is a collegiate graduate.

Found in the person of another patient, Emery Owen, of Fairfield, Ohio, a Good Templar brother.

To-day, upon returning from forenoon service, found the expected telegram. I take the early morning train for Louisville.

CHAPTER

JEFFERSON HOSPITAL,
JEFFERSONVILLE, IND., October 18, 1864.

This large, new hospital is located on a bend of the Ohio river, just across from Louisville.

It is built on the plan of the "Pavilion," like the Chestnut Hill hospital, of Philadelphia. The sick wards are of one story, twenty-four in number, and radiate from a circular covered corridor, like the spokes of a wheel. This circular corridor is half a mile in extent. and fifteen feet in width, enclosed upon the sides, and provided with windows and doors. Within the circle are the buildings of the executive department, rooms of surgeons, full and light diet kitchens, dispensary, dead-house, post-office, printing-office and chapel. Crossing this circle and leading to these central buildings are two covered corridors which cross each other in the centre at right angles.

Each ward is one hundred and fifty feet in length by twenty-two in width, and contains fifty-nine beds for patients. To the rear of each ward, is attached one small room for wardmaster, another for clothing, besides a bath-room and closet. In front of each ward, is attached a little dining-room and pantry. In the latter place the diet is dealt out for the patients. This is brought hot from the kitchens, in covered tin cans, in a little hand-cart on wheels, upon which is marked the number of the ward.

Thus one might live here for months and not go out from under cover, be very hard at work, and walk several miles each day. That cleanliness is essential to health, seems to be a prominent idea, and the wards and corridors undergo a scrubbing twice a week, and mopping as often besides, which gives a neat and wholesome air throughout. Upon the arrival of patients they are disrobed of their dusty if not filthy clothing, it is rolled up, a check given for it and it is packed away in the baggage-room, together with their arms, if any. They are provided with clean hospital clothing and a clean bed, which is changed each week.

The laundry is a building separate, and some distance from the hospital, upon the immediate bank of the river. This is supplied with some twenty or thirty large, bare-armed water Deities, who probably swam over from the emerald isle.

One of the wards contains the large dining-hall for the ward-masters, nurses and guards, a smaller one for the stewards' mess, and opposite, another for the ladies' mess. Above these are sleeping-rooms, two of which are occupied by the lady nurses. It is in contemplation to supply each of the twenty-four wards with one of these last-named dignitaries. A few are without, some to their professed grief and vexation of spirit.

SATURDAY, 22.

The afternoon of my arrival, attended funeral service in ward 23, of Private Isaiah Lusby, Co. I. 9th Ohio Cavalry. Chaplain Fitch, of the regular service, and former tutor to Secretary Stanton, spoke well and briefly from the words, " As much as *lieth in you* live peaceably with all men." Language implied that a man might burn and rob property for you and it would not lie in you to live peaceably with him, and that a good and just Government might have rebellious subjects.

We have no stove in our sleeping or dining-rooms, and

really suffer with cold. We occupy the single, iron, army bedsteads with hard husk beds; but these discomforts are doubly counterbalanced by the pleasure of ministering to the comfort of the sick boys.

This P. M. rode over to Clay Hospital, branch C., expecting to take the place of the matron Miss Wells, while she is absent on a thirty days furlough. But as she was going ostensibly to take an invalid soldier to his home in Michigan. and as all of the Michigan boys are going home to vote, some twenty from that hospital, the surgeon says that twenty can take care of one, and her services are needed in the wards. Returned, and am assigned to duty to Ward 1.

One lady came here a few days since, who staid only two days. She was " not used to any such fare, such cold rooms, and couldn't work for any such pay." There are others here who do not work for the " pay," but for something higher and better.

Tuesday, 25.

Have been learning of my duties, and getting acquainted with patients. On Sunday eve had singing in my ward. Mrs. Rhodes of the gangrene ward, Mr. Wheeler, and some four or five of the convalescents sang " Homeward Bound," " Oh. Sing to me of Heaven," " Rest for the Weary," " Shining Shore," and " Rock of Ages," to the evident and warmly-expressed gratification of all. Think we shall try to inaugurate the practice in other wards, it seems to do the boys so much good.

On Monday, one was taken from my ward to the gangrene tents. His arm was in a bad condition from impure vaccination, and now the gangrene has appeared. It is said to be worse than a wound to heal. Three of the worst patients

21

complained of wounds smelling so badly that it kept them nauseated. Procured the last couple of handkerchiefs from sanitary stores, and a piece of old muslin which I hemmed, and saturated all with cologne, which had kindly been donated for the purpose, by " Gale Brothers," of Chicago. Two received them with simply thanks and smiles, but the third, a Pennsylvanian exclaimed, enthusiastically : "Oh! my gracious, now if that ain't nice. You couldn't please me better than to bring that there, it 'ill kill all the smell sure, of my arm. I allays was sich a feller for cologne and hair ile and all sorts of scentin' stuff, when I was to home !"

At present there are thirty-seven patients in my ward, twenty-three of whom are wounded. There is but one just now who it is thought will not recover. He was shot through the upper portion of left lung, has a bad cough, no appetite and is emaciated. His parents live only about eight miles distant, on the Kentucky side. We call him Willie.

There are no others who cannot sit up, if only for a short time, while the majority are able to do so considerable, and to walk about. Still, were almost any of them at a northern home, and transferred into " our boy," or " my husband," he would enlist the care and sympathy of a neighborhood ; and justly so, unless the kindness should have the effect it did on our dining-room boy, who says he was never sick until since he went home, after being in the service three years, when he " ate himself sick."

Upon first entering the ward, after being assigned to duty, found one man who was bitter against red tape, nurses and surgeons in general, and his own in particular. Said he : " Didn't have anything fit to eat, guessed the nurses got it all, the doctor was as mean as he could be, and hadn't been near him for two days."

I found that he was excited and half-crazed with the chills, and hope deferred. He said he had been in the service three years, that Government is owing him $232, but could not get his descriptive roll to draw it, as his captain was lying at the point of death. His wife was needing money, and he wanted to get that which by a recent law is due him, without descriptive roll, viz : $32, and he also wanted a transfer to Cincinnati, Ohio, but he " Didn't expect to get anything, Government had got the service out of him, and that was all it cared about, nobody cares for me. I'm *only a private.*"

Soothed him by promising all I could do, excused remissness in others, from over care and work, and promised to intercede for a transfer to Ohio, and his pay. The next morning, while dealing out diet in the little pantry, his plate was sent back *full.* Upon going to him, to find out the trouble, as the food was such as I had ordered at his request, light, warm corn-bread, butter, eggs, fruit and coffee, found him sitting in a sullen mood by the stove, and would neither have that nor anything else, had " eaten all he wanted." The surgeon came in, soon after, and told him he had put his name down for a transfer to Ohio. I learn that he has been so near deranged, that one night, not long since, he jumped out of the window, ran to Ward 2, and reported that they abused him so he could not stay. Poor fellow, home is the medicine for him.

TUESDAY, November 1.

Visited gangrene tents to find four patients from my ward, who say they wish to be considered patients of Ward 1, and shall expect to be looked after, occasionally, by the lady matron of said ward. I never saw or scarcely imagined such suffering as the poor fellows undergo from the application of bromine, and do not wonder they have christened the place " purgatory."

It will be necessary to imbibe a little more of the heroic,

before I can be of much help during an operation. The
red and swollen elbow of the arm which may yet fall a victim to impure vaccination, was resting in my hand, while the
nurse proceeded to take off the oakum which had been saturated with bromine, and then to pick off from the side of the
raw wound, the burnt pieces of flesh, with a pair of pincers.
I could have seen this done if it had not hurt anybody, but
when the sick man began to cry for mercy and his elbow
quivered in my palm, everything began to grow strangely
dark, and knowing from past experience, that they might
have another patient to care for, in a moment more, I dropped the arm into the hand of Mrs. R. and mentally calling
upon the heroism of all the braves I had ever heard, reeled
to the tent opening, pulled back the curtain, and in a moment
things grew lighter. All laughed at me, even to the patient;
but it isn't to be expected that a Yankee school-ma'am can be
transformed into a dissecting surgeon in a minute, guess it
will take about a fortnight.

At the request of patients, had a sing in the tents that eve.
On Sunday attended funeral service of a soldier by the name
of Rogers. In the evening, attended church service in town.

EVENING.

Just after tea, the following letter was sent to me from one
of the patients, addressed to " Lady Matron, Ward 1."

Miss: I was informed by a gentleman last eve. that you reported me as being drunk and boisterous. Is it possible that a
lady of your qualifications, capable of adorning the best of
society, can so far forget herself, as to report one for such an
offence, without even admonishing him of the wrong he has
committed, and to what it will lead, if followed up ? Perhaps
that one has a lovely wife, the companion of his childhood,
and now linked to him by closer ties, with all her future hap-

piness depending on the character of her husband? Would you knowingly mar her happiness without even raising a warning voice to the one to whom she has risked her all? Oh! I cannot believe that you would be so cruel to one you never saw. or to one you have seen. There must be a mistake somewhere, hence you will excuse me for taking this mode of asking you, not in my behalf, but in the behalf of those friends that are near and dear to me. Please inform me of the truth of the matter.

Yours, in haste,

(*Signed*) ————."

Have written the following reply which will be lain upon my little stand, in the morning, where the other letters are placed, and where he will find it, though there are so few called by their right names here, that I havn't the least idea who he is :

"Mr. ————, Dear Sir, I received a note from you last eve, in which you say you were informed by a gentleman that on Sunday evening, I reported you at head-quarters for drunkenness and boisterous conduct.

It is all a mistake. I have reported no one, neither have caused any one to be reported. It would be necessary to know the person by name, before he could be so reported; and the only one I have even *suspected* of having drank too much, in my ward, is one whose name I do not know. That person may or may not be yourself; but it has not, by me been so reported. If it was done by any one, probably some man has done it, who like Adam, was not noble enough to take any part of the responsibility upon himself, but like him could say :

'The *woman* that was given to be with us, *she* did it?' But Sir, you do not deny the fact of being in that condition;

22

and perhaps the one who reported you, if indeed you have been reported, which I doubt, considered it a duty, and it might have been.

You say I ought first to have raised a warning voice to you for the sake of that wife. Let me do this now. You are still sensitive—still careful of your reputation for the sake of that ' dear wife.' Let me beseech you as a friend to abstain entirely and at once from the use of liquor in whatever form. Look not upon the red of the wine-cup. Be a strong, noble man—strong to overcome the temptation, nobly battling against it, that if you conquer you may be ' greater than he who taketh a city.' I am a member of the Order of Good Templars,—therefore the more interested for you. If at any time you wish any advice or sympathy in my power to give, *while battling against this sin,* do not hesitate to speak or write to me. In conclusion, after beseeching you not to entertain the idea that a soldier's life necessarily calls for liquors, I will quote those beautiful lines for you of Dr. Holland's, on the subject of Temptation:

> ' God loves not sin, nor I,
> But in the throng of evils which assail us,
> There are none which yield their strength
> To Virtue's struggling arm, with such munificent reward of power
> As great Temptations. We may win by toil, endurance ;
> Saintly fortitude by pain ; by sickness, patience ;
> Faith and trust by fear : but the great stimulus which spurs to life
> And crowds to generous development,
> Each chastened power and passion of the soul,
> Is the *Temptation* of the *soul to sin,*
> *Resisted and reconquered, evermore.'*

Yours for reformation,

(*Signed*)   ———."

WEDNESDAY, 16.

On Saturday evening a printed order was sent to each of

the wards, that the "surgeons thereof must send in the diet lists each morning, in their own hand-writing, as it was feared in some instances the lady nurses were allowed to make them out."

The next morning the ward surgeon *copied* the one I had prepared and sent it in his own hand-writing. I am making out "Morning Reports," also of number of patients—sick or wounded—from what hospitals, &c. Had just finished one on Monday eve when seven new patients arrived. Made out new one, when the surgeon told me to copy his signature and sign it, which I did. Yesterday morning Dr. C. made out the diet list, and put two of the worst patients on *full* diet. Finally, after convincing him of the fact, he asked me to sit down and he wrote while I dictated, thus complying with the letter of the order, while the spirit was best carried out.

Several wounded men who are obliged to lie in bed all the time, have been for some time sadly in want of hair matrasses. There are a plenty in the store-room, but they have refused the request of the ward master and nurse, even with the order of the surgeon. They say they are keeping them against the arrival of other patients. But four of my boys were suffering so much for them I obtained the order from the surgeon, went to the store-room and left the order, and a request that I might be allowed to send down my own mattrass—which one of the elder ladies had managed to obtain for four of us—in case we could not get those at the store-room, to the surgeon-in-charge Dr. Goldsmith. The clerk would not let them go, without first seeing him, but in about two hours the clerk came over to give the ward-master a piece of his mind for letting a *woman* interfere in the matter. But he was reminded that the trial had been made by himself to no purpose, and I didn't care how he felt about it when I saw the smiling faces and heard the warm expressions of the poor sufferers, when they were moved on their nice, soft beds.

Nov. 17, in Ward 1.

I steal a few moments to write, while surrounded by patients who are walking, talking, asking questions, etc., which certainly does not have a beneficial effect upon composition.

We had, this morning, every bed full—thirty-nine sick and twenty wounded. But since then have had a fresh arrival of several hundred patients at the hospital; consequently those who have been detailed for nurses or attendants in my ward must give up their beds and sleep in tents. This is all right —I am glad to have the ward filled up again. During the furlough to vote we had but seventeen patients, and now have but three, who are too sick to sit up a part of the time at least. But there are some four or five others whose wounds oblige them to lie in bed. Willie's appetite is better and we hope he is really getting well.

Beside duties previously mentioned I have been engaged in others. I have charge of the diet—assist each meal in dealing it out. I have covered crutches, ripped up arm slings, washed and made them over, gone to commissary with order from doctor for material for pads for wounded or amputated limbs, and manufactured the same. I *petition*, and thus commence the transfer or furlough of one or the pay of another. I write letters for my patients, read or sing for them, visit or play checkers with them, occasionally, to make them think they are at home and forget they are sick. Have once, through the kindness of the one detailed as baker, been allowed to make some cake as a treat, in which the patients of Ward 2 and the gangrene tents participated. We have a sing in the wards about twice each week. The convalescents are invited from adjoining wards and we have quite a crowd and pleasant time. Every ward is eager for its sing. I have also bought some cheap prints, put on moss frames, arranged a wreath of autumn leaves on white paper, and have tried to have something on a little stand, which should represent or bring to mind a cabinet, to make them think of home. In short, have tried to make my ward look as Miss B. expressed it, " as if there was a woman in it."

The surgeon, ward-master and nurses treat me with the greatest respect and consideration, as well as the patients, and I am certain the latter appreciate the little I am able to do for them.

But the bugle has just blown for the carts to start for the kitchen—they will soon return—mine first, and I must hasten to the little pantry to deal out the supper for the sick and wounded boys.

SATURDAY EVE, 19.

My writing progresses slowly of late and is often interrupted, for I am very busy. I would like to note down the duties and incidents of *one day* if time permitted, but can only select a portion.

Day before yesterday was gladdened by a call from Rev. H. M. Miller, Agent of Universalist Army Mission and his travelling brother, Rev. Gilman, Michigan Agent. I regret that he cannot be allowed to preach in this hospital. This narrowness of religious thought reminds me of the early history of an own father, long since sleeping in a western wildwood, who when a young man was repeatedly denounced from the pulpit of a Baptist divine, who cautioned his hearers to beware of the fascinations of that Methodist fanatic, who was setting the people crazy with his preaching. Am wondering how many years it will be before people *can* worship God according to the dictates of their own consciences, with none to molest. How many before Universalist papers can be given out as well as Methodist ones to sick men who prefer them,

instead of being carefully collected and torn up or burned by
those who think they are doing God service. What a pity
that so few who fight for civil liberty know so little of religious freedom. But such is humanity—boastful of God-given
rights, freedom and equality, while in blissful ignorance of
their own manacles.

We are expecting a Thanksgiving dinner at the hospital
next Thursday, for the setting on foot of which we are indebted to the efforts of our kind Chaplain Fitch. But as so
many citizens in Jeffersonville and Louisville are not any too
loyal, feel somewhat dubious about the turkeys, chickens and
pies for two thousand mouths. Certain it is that the boys
would appreciate a good dinner, as they have had rather short
rations of late, and there has been some just grumbling by
the full diet patients. And yet it is in most things a model
hospital, but must be very difficult to supply so large a moving
population.

Often, I see the time, when if I had a box of sanitary
goods, the patients could be made more comfortable. It might
be different for one to understand why this should be needed
in a hospital of such resources as this, and will note a few
instances.

At one time I found a man in the gangrene tents who had
not had a clean shirt since he had a hand amputated five days
before. The garment was spotted and stiff with blood of
course, and he had repeatedly asked for one, but had been
told clothes were issued but once a week, Had I a box containing such an article he should not have waited an hour before
having one; as it was he did wait a week. The ward-master could have drawn one by obtaining an order from the
doctor.

Two men in my ward having wounded shoulders could get
but one sleeve on, while if I could have obtained those with

open sleeves, tied with tape, it would have been more comfortable for them, besides presenting a better appearance when sitting up. Two, at another time, could think of nothing they could eat, except toast and canned peaches or other canned fruit; but although I obtained an order from the surgeon immediately, I could not obtain the fruit, as it was not in the sanitary stores for a week afterward.

Other instances in which I could have made good use of a box from an Aid Society have occurred several times, and to-day when a man needed a pair of woolen socks. We have been informed for the past two weeks that it was of no use to make out a requisition for them, as they have none to issue. Some three pairs for the most needy have been given me by a lady nurse recently from the North—a contribution from an Aid Society. For a time also, we were destitute of handkerchiefs and now no ginger wine can be procured. Sometimes a poor boy thinks if he only had a little butter which came from the North, and was not so rancid as what we have here, he could eat something.

A couple of gentlemen have just come in with a note-book, and we have been singing for the invalids. It is getting somewhat late to be in the ward—about eight, and I must close this rambling memoranda for this time.

MONDAY. 21.

Yesterday witnessed a Sunday morning inspection for the first time in our ward. The bugle sounded, the ward-master took his position by the open door, each patient who was able to sit up took his place by the side of his bed, and the nurses and attendants ranged themselves upon each side near the door. When the surgeons appeared, at the word of "Attention!" from the ward-master, each man rose to his feet who was sitting but able to stand, and the Inspectors marched swiftly

through the ward to the bath-room and back through the
ward, pausing only to compliment the ward-master upon the
"usual fine appearance of Ward 1."

The corps of inspectors varies on different mornings, but
this time we were honored by the presence of the surgeon-in-
charge, Dr. M. Goldsmith, the executive officer, the officer of
the day and our own ward surgeon. The first sported the
gold leaf of a major, the officer of the day the green sash, and
all the gilt stripes and buttons of the medical department, and
our surgeon the U. S. upon the shoulder. I was uncertain
about what should be my own position, having thought nothing
about it. I was reading to my sickest man, who was lying in
bed, and rose to my feet also to receive our guests but sat
down before they returned from the bath-room. Was honored by a lofty bow from two or three of the dignitaries.
Determined to know whether I ought to rise or not before
next inspection day and referred the matter to the surgeon,
who said:

"It is the soldiers who are expected to rise and you are not
a soldier, are you?"

That settled the matter, the dignified matron could here-
after sit in the presence of her betters.

Wrote four letters to-day for sick men and have commenced
the transfer papers of Frank N. Button to Detroit, Michigan.
He is a young boy—has been here five months, and is a quiet,
patient sufferer. His left limb is paralyzed from a wound in
the hip, and I fear will always be useless. He has not stood
on his feet in that time except as he is held up. I have written
for a friend of his to come for him. A sing in Ward 1
to-night.

FRIDAY, 25.

Well, our Thanksgiving dinner was a success. Nearly

three hundred turkeys and chickens suffered death for the good of their country. When those, and the five hundred pies were cooked and placed on the tables in the large, full-diet kitchen the night before, I mentally confessed, while viewing them through the window from the corridor, that were I one of a regiment of hungry soldiers just from the front, I might possibly stir up a mutiny to make a raid on the kitchen and capture them. A portion of the dinner was the contribution of the loyal citizens, and about one-third was furnished from the hospital fund.

The chaplain sent for me as usual to attend funeral service. To-day it was in Ward 15, and of four soldiers. One was that of George W. Odell, 28th Michigan. He was but seventeen, in a new regiment and only out about four weeks. He had an escort of eight young boys of his company who appeared in uniform, with white gloves and reversed arms. We ladies followed next to the coffins in the procession to the ambulance. The latter conveys them to the soldiers' cemetery.

It is with us only "a funeral service" of "one, two, three or four," as the case may be, "in" such a "ward." The forward coffin bears the stars and stripes. A short Episcopal service is held, and we follow to the ambulance. But we know, though fast learning to ponder less upon it, that somewhere is one more vacant chair, and missing voice and footstep, for every death which occurs here, and sorrowing hearts, to whom a few words of condolence and a lock of hair, sent by some matron, or the official blank properly filled out by the chaplain, comes almost as a mockery in place of the dear boy, or husband, or father, who left them with such vigorous health and bearing but a little time ago.

> "And yet, and yet, we cannot forget
> That many brave boys must fall."

23

But we comfort ourselves with the thought that though

"Their swords do rust, and their steeds are dust,
Their souls are with the saints, we trust."

MONDAY EVE, 28.

Saturday eve had singing in my ward. Benches were carried in. Chaplain's orderly, Mr. Bullard, brought in and
distributed as usual the little army hymn books. Patients
were invited in from other wards ; we had quite a crowd, and
a pleasant time. Our ward surgeon was also present, with
the usual singers, viz., Corporal Patten, Steward Holt, and
Burroughs, Wheeler, Dupont, Artillery, Perry, Payne, and
ladies Dixon, Lawson, Hardy, Rhodes and Sturgis.

Yesterday was very busy all day in ward, with new arrival
of patients from Nashville. Did not get time to attend service. Have also been very busy to-day with same. Have
written out applications for transfer, filled out medical descriptive lists, except the diagnosis, and have written out orders for
money to be paid to the surgeon for patients unable to get to
headquarters. We ladies signed the pay-roll yesterday morning. The clerk had by mistake got my first name wrong and
had to sign it the same. Easy way of changing one's name.

We have one singular individual who goes by the title of
" Colonel." He came with the transfer of patients from
Nashville, which consigned ten to our ward, two weeks ago
last Wednesday.

He was brought in on the shelf which was taken out of the
ambulance and placed hastily upon the bed, while the nurses
hurried out for more. They had lain his head below the
pillow instead of on it, and seeing him lie thus without raising
it, though he appeared to make some ineffectual attempts to
do so, I went to him to assist, and asked if he could not raise
himself higher and on the pillow. He said no, that his limbs

were all paralyzed except one arm. He raised his head and I put the pillow under it, and when the patients were all brought in had the nurses lift the man up higher in the bed. Soon after, when accompanying the surgeon, while he was making out the cards to hang in the little tin case at the head of each bed, the patient informed him in a confidential tone that he wanted his name entered as a private, as the boys were always jealous of an officer and expecting him to put on airs. But that he was colonel of an Illinois regiment. Also that he had been robbed of his satchel, clothing, regimentals and $3700 by the ward-master of Ward 1, Hospital No. 8, of Nashville.

He is looked upon by the surgeon and others either as an imposter who is trying to "play off," as they style it, or as crazed from the effects of fever. I have preferred to give the latter more charitable verdict till I know the opposite, and in spite of some opposition have treated him accordingly. His appetite has been perfectly ravenous, and beside supplying him with the rations of two or three men each meal I have bought him apples and cake to give between the meals, with money given for that purpose by his brother, who has been down to visit him. I was at first fearful to give him so much and did not until he cried and begged for it, and I found it did not seem to hurt him. Three men's rations for the day, lately are nothing, he wants and gets about six.

After he had been in the ward several days and been lifted about by the nurses, as though helpless as a babe, it was confidently told me by the ward-master, chief nurse and others, that it was their belief his paralysis was mere pretence. He had been teasing me to intercede for him to get a furlough, and the next time I saw him he repeated the request, when I informed him that no furloughs were given to such as were not able to walk to an ambulance or step into a car, and tha as soon as he was able to walk about, I would try to get a furlough for him. That I wished him to get up and be dressed that afternoon, and sit up a while, and do so each day, and try to use his limbs and perhaps he might get the use of them. Told him that I would come in the ward in about an hour and bring some work to sit awhile, and hoped to find him sitting in the rocking-chair. Went at the time and found him sitting in it and looking rather foolish, and I fancied then, as from the first, that his eyes looked as if he had been imposing upon our credulity, but preferred to give him the benefit of the doubt and think him half crazed. He then paid some silly compliments about ladies' society and wished me to sit near enough so that he might rest his feet on my chair—"they were weak yet." Asked him if he thought he could raise them, and found that he had walked from his bed to the chair. There were many others near and who heard the request, and after some hesitation I preferred treating him like a sick child, and turned the chair so that he could put his feet upon the side rounds. The next day he sat up again at my request, and upon the next, when entering the ward, found the *paralyzed limbs performing a shuffle* to accompany a tune he was humming. I expressed my satisfaction that he was improving so rapidly and prophesied a furlough. I was half tempted to prophecy instead a return to the front, which would no doubt have taken all the strength away, and beside I really thought his mind was not right and perhaps a visit home might restore him.

I had obtained the consent of the doctor to put his name down in the next furloughs which were granted, when last Saturday he became angry with the nurse who had ordered him to use the spittoon instead of floor, and ran away to head-quarters. Said he wouldn't stay there any longer and wanted to be sent to another ward. While I was away he was transferred to Ward 6. I visited him there yesterday, and found as I had expected he would be, as soon as recovered from his anger, very repentant, and sorrowful that he was there, saying childishly, " No other ward will ever seem so like home— there's no lady here, but whenever I wake up I fancy I hear your step bringing me some apples. Won't you ask the doctor if I can come back? " I promised to do so, for the poor fellow was shedding tears, but the doctor says he ought to stay there for being so foolish.

FRIDAY EVE, December 2.

Have this eve parted with Frank—the patient mentioned under date of November 21. His mother and a gentleman both came for him, but unknown to the other. Their expressions of gratitude at parting, which seemed extravagant, have done me good. I am hearing too many blessings now-a-days from sick and dying men to be in doubt any longer whether or not I am doing good.

Yesterday felt very sad that one of the patients who desired to get a transfer to Mound City, Iowa, near his home, was instead sent to Madison, Indiana. Had I known of the intention before the name was sent to headquarters, or had the surgeon not forgotten about the transfer through the multiplicity of his duties, it would not have been. This was one trial, but the worst was the transfer of the " colonel" at the same time. It was too bad. I petitioned the doctor of Ward 6 in vain. Have written his brother where to find him, and supplied the " colonel" with paper and an envelope addressed to myself, and he has promised to write to what ward he is taken, of which I shall inform his brother. I still think him half crazed from the effects of fever.

Last Wednesday eve occurred the very pleasant little incident in my ward of the presentation of a gold-headed cane

24

and gold pen from the patients to our ward surgeon—Dr. J. M. Chapman. A nice little speech was made by our worthy Mr. Bayne, of Philadelphia, and a very happy impromptu reply from the doctor.

FRIDAY, Dec. 9.

The first snow of the season. Winter has really come to the Ohio valley.

Much public excitement in Louisville. Men are being conscripted, and horses impressed. Several thousand soldiers have just been sent there, as they anticipate a cavalry raid from the rebels. Hood is threatening Nashville. He says he " is ordered either to go into Nashville, or to ——— " a certain very warm place. Our boys think he will get into the latter place first.

One night last week, a man in an adjoining room had the nightmare and woke us all up three times. At the last, he was taken to the guard-house. The truth was he was intoxicated, and it was also the third offence. He was sent to the front next day, as is usual. But he was not, as was laughingly reported, put in the guard-house and sent to the front for having the nightmare.

Yesterday was at work most of the day and evening on evergreen wreaths to trim the ward. Christmas is coming! I have plenty of help from the ward-master, chief nurse and convalescents. How kind they all are. I receive nothing in my ward from the surgeon down, but the greatest respect and consideration. Some of the ladies can get no assistance, but those in our ward are ready at all times to help.

FRIDAY, 16.

The first death in my ward, since my coming, occurred last night. It was that of Robert Burnett, of Kentucky. On

Sunday morning, over a week since, I found him lying in bed and that he had not been out to breakfast, as he had done the two days previous, since entering the ward.

Upon conversing with him he told me he was going to die. I saw that he was excited and thought he was nervous and tried to quiet him. But he was sure, he said, that he should die, " he understood why I did not think so, and appreciated what I said, but he *knew* he was going to die, "and asked if I would stay by him whenever I could, and he begged for a promise that I would be by him and "watch his face when he died." These were his exact words, and though I did not think he was dangerous and told him so, yet he would not be pacified till I promised if he died at any hour when we were allowed in the ward, or if at any other, and he was conscious and would send for me, I would be with him. He was also concerned for the future, for he was not a Christian, he said. I read for him from the Bible, sang for him, and the chaplain's orderly came and prayed with him. He professed afterward to think himself prepared to die, and he gradually grew worse each day until he died. I remained with him until late last evening, but he was unconscious else I should have remained until his death. He died about twelve. I had written to his wife the first day, but the mails are interrupted by guerrilas. He has two brother-in-laws here, who have started home with his body. At the funeral service we sang the appropriate hymn,

> " Oh ! watch my dying face,
> When I am called to die."

WEDNESDAY, 21.

Transfers and furloughs are the order of the day. Some twenty-five hundred have been transferred from Nashville to this hospital, this month. From fifty to two, three or four hundred are transferred from here at one time, to hospitals farther north. As we hear that those are pretty well filled, it seems just the time to give as many sick furloughs as possible, thus clearing the hospitals for those unable to go home. I will give a sketch of one who has just gone home on a sick furlough.

His name is King, and his home is in Beattyville, Kentucky. He came here from the hospital at Nashville, about six weeks since. He had suffered from extreme exposure and hard marches which had broken him down and induced fever. Gradually, slowly the coaxed appetite returned, the mind recovered its tone, for he had sometimes fancied himself a major, at another he had met me in the morning, with an anxious, puzzled expression and inquired if I had seen that man to whom he had given his money. "No, I had not." For some moments I fancied that some unscrupulous person had been taking advantage of his illness and recent arrival, and had inveigled him into an unwise consignment of money, particularly as he told me the man had said he kept a safe for keeping soldiers' money.

But upon further conversation, in which he averred that "the box of money had been sent to him, and part was in gold," his delusion was manifest. But unlike some, he was easily made to understand that fact, and like a hero he strove against such phantasms. About three weeks since, he received a letter from his wife, which he brought me one morning to read.

It was difficult to decipher, even for a Yankee schoolma'am, from the peculiarly original style of orthography and of punctuation ; but Yankee ingenuity triumphed, and revealed a volume of suffering. The pages were eloquent with starvation, affection and loyalty.

"Come home dear Dick," was the burden of the letter, "or we shall starve. I have but the milk of the one cow for my-

self and the four little ones to live on. And the cow gives
but the half-gallon a day. The guerrilas have been in here
and robbed the union folks. You tell me to go to Mrs. H. if
I need help. She has made friends with the rebs to save her
property, and I'll starve before I'll ask her for help."

Though I started out immediately on a scouting expedition,
in search of something available to meet the case, I almost
envied one of the other lady nurses, who, had received $60,
but a little time before, entrusted to her for the soldiers
from an aid society. I would not ask her to assist me, for she
would have need of the money in her own ward, but the generous Chaplain Fitch had called me " daughter," times enough
to encourage a demand upon his generosity, at least I would
read him the letter and have him converse with the man
whose manners and words impressed me with a belief in his
honesty. The chaplain came, heard the story and letter, and
placed a " V " in my hand at starting, to send the starving,
patriotic woman. It *was sent* not confiscated. Whether received or not, I cannot at present say, but hope to know in a
few days.

About the same time a furlough was requested for Mr. King.
The time went slowly by, at last the furlough came. The
poor fellow had no overcoat, no haversack, no money, having
lost these articles when taken sick during a forced march.
By a late order of the war department, the absence of his
descriptive roll, for which I had long before written to his
captain in vain, prevented his drawing any clothing. Flannel I had previously obtained for him from the sanitary commission. But their stores contained no other needed articles
of clothing. It was stinging cold, and he must go warmly
clothed. The ward surgeon sent the man over to tell his
plain story to the executive officers. But stringent orders
must be obeyed, and he came back empty handed. I went

next to Miss Buckel, then to the chaplain, and after some delay the man returned with an overcoat. The ward-master captured somewhere a pair of shoes, and a haversack of white drilling, into which he put four days' rations of bread, meat, sugar and coffee.

Now where was the four dollars to come from, which he would need the last part of his journey to pay stage fare? So far as my own private purse was concerned, I had received just that amount at the last pay day, and had scarcely any beside to last for the next three months. I had the idea—I would write a note explanatory of the case and of my belief in the worthiness of the man, and state the principal object of his going home that of getting his starving family out of rebeldom. I would address the note to a friend of the Sanitary Commission at Louisville in particular, and to all generous, loyal people in general. I told him if he lacked friends while on the route or got into trouble, if there was a loyal person to be seen, especially a Sanitary Agent, to present the note, it might do him some good, and would certainly do no harm unless he might fall among rebels.

He started, and at night he returned word: " Tell Miss P. that the order she gave me did me a *heap* of good. Tell her it got me a nice pair of woollen mittens, a great long piece of tobacco, four dollars in money, and a note from Mr. Scott to a man, where I'll be to-night, to give me a good supper and see me started on my way in the morning. And more and better than all, he has given me a note to the captain in the nearest regiment, who will help me get my family away."

So my little note seemed likely to be an " open sesame " to him everywhere. Wasn't I glad ? Didn't that pay for getting up at an early morning reveille, standing on an icy floor by gas-light and handling dishes which are frozen together, while dealing out the diet? Of course it did! And this is not a solitary instance of my reward either. The wanderer has promised to write to me ; and when he returns, if not before, shall know something more of his journey.

SATURDAY, 24.

The second death in the ward. It was that of a young, noble-looking man—Prevo, of the 40th Indiana. He died of a gunshot wound, the ball entering the lungs. He was battling with the grim monster all day yesterday, and thought himself at one time on a forced march through the country of an enemy, and at another in the heat of battle, when he would cheer on the soldiers. A lock of hair and a few words of condolence will go to one more mourning family in place of the dear, noble boy.

Great preparations are being made for Christmas to-morrow ; thus death and feasting go hand in hand in this strange world of ours.

Another died last Sunday in Ward 23, who had been for a long time in this ward. He shed tears when he was transferred, and I interceded to have him remain, but there are wards to which an order obliges patients to be removed when suffering from chronic diarrhea or lung diseases, and he was one of the former. But at his request I visited him, and after his death, which came suddenly, procured a lock of his hair from the dead-house and sent it to his father.

CHRISTMAS EVENING.

Our dinner was truly a success. It was given by the Sanitary Commission principally, and a portion from the hospital fund. Much less stir was made about it, and one soldier expressed the general feeling, who said he " enjoyed the Christmas dinner the most, for there wasn't so much style about it." Very excellent oyster soup for the light diet was given each time. Twenty-one hundred pies were issued for dinner, seventy-one cans of oysters, with eighteen hundred pounds of beef *a la mode*, also four barrels of pickles.

But this must have seemed so like a mockery to one mourning wife who is here. Sergeant Don A. Clark, a very worthy man and Christian, who, Chaplain Fitch says, "has suffered more than any other two men ever in this hospital," died just after midnight. He belonged in my ward, but when I came here he had been sent out to the gangrene tents. The ball had passed through the limb a little distance above the ankle but had injured no bone. It was simply a flesh wound, and little trouble was anticipated in the healing. But after sometime his blood was ascertained to be in a poor condition, as indicated by an eruption upon the face. This is considered a bad omen when a wound has become inoculated with gangrene.

He came back to the ward once, for a visit, on crutches. He was hopeful for "the surgeon had told him he would soon be well enough to go to his own ward," and boys "said he, I shall be back home with you in a few days."

The wound has several times been free from gangrene, but just as he was anticipating a return to the ward it would return.

Thus did he suffer with hope deferred and the cruel burnings of bromine, as could only a noble, patient christian, till from two little wounds the size of a hickory nut, it extended to nearly the whole lower limb. It was shocking to see the the cruel ravages of the gangrene.

Then it ate off an artery, and twice he came near dying with hemorrhage. Then it was hoped he would rally so that they dare amputate the limb. His wife had been all the time writing to him for permission to come and care for him, but he had been hopeful of going to her, and the expense was an item. But he was at this time prevailed upon to give his consent, and I wrote to her just what we hoped and feared· That she must expect to find a sufferer, but if she could come with nerve and moral courage enough to hide her feelings in his presence, and smooth his passage to the better land, we should be glad to have her come, and the expense should be nothing during her stay.

She came about three weeks since, and has proved herself equal to the task. His mind became very weak, and once when I carried him some currant shrub, he cried like a sick child as he said, " God bless you." I went in last evening to see him for the last time. He had forgotten almost every one but his wife, and as I took his hand he appeared not to recog nize me, even after I had given my name, but as I mentioned my ward, he said earnestly, and with tears and a tender childish voice, " Oh, yes, I know the lady of Ward 1. I never *can* forget her, she has done a great deal for me."

Such is my reward. Gold, without it, is as nothing in comparison.

FRIDAY, 30.

Most of the wards are now radiant with evergreen, tissue paper and pictures. I am content that mine should rank third or fourth in its adornings, rather than neglect the weightier matter of attending to the sick men—of whom I had quite a number last week requiring much care. The last death, mentioned under date of the 24th, was the second only in the ward since my entrance—a period of over two months, and the fifth since being in the charge of the present surgeon, which is eight months. But the mortality in the hospital is increasing very much in consequence of war's grim visage approaching nearer to us. A week ago last Sunday there

25

were eleven dead bodies in the dead-house, and fourteen deaths occurred in three days.

Last evening I was edified by the testimony of a loyal Kentucky woman who is visiting her wounded son in one of the wards. She said:—

" Well, I'se never in a free State afore, but I've been very much gratified to see how our soldier boys is took keer of. Talk about the *riches o' Kaintucky*—I say the *riches o' the North.* I wish every one o' the rebels in Kaintucky could see what I've seen here." And here the old lady, with her good motherly face surrounded by the full border of her cap, and the forefinger of the right hand brought down with three or four emphatic taps into the left palm, forcibly reminded me of Mrs. Partington, as she concluded her observations thus : " They'd just be like a pea fowl when it drops its tail feathers to the ground,—and they'd never cry ' Yankee ' any more."

Have lately been the recipient of what I presume was intended by the writer to constitute the first of a series of love letters addressed to " Mr. E. J. Powers, Esq." I had addressed a note of inquiry relative to a boy who had died in my ward, to one whose name he gave, inquiring for his sister, signing only last name with initials of first. The letter, without telling me a word of what I wished to know about the whereabouts of the dead boy's sister, contained the following delectable bit of composition:—

" Mr. E. J. Powers,

Dear Sir,—Please excuse me for answering you likewise. I must confess that I was infinitely pleased for you to write to me and inform me of the mishap. (The death.)

Soldiering must be a dreary life, altho' I have never experienced it of corse because I am not a masculine, but woo to me if I were.

I have nothing more to say at present, but by the way if you have no objections I will compose for you a little piece of poetry.

" think not this a hapy
world for it is ought
but a world of care
and trouble.

remember thou art
but a mere vision
yet we think not so
For our body seems like
if we could overcome
every thing but! Ah ! it
is all in vain."

Signed ———.”

Only to think what an amount of poetical talent will waste its " sweetness on the desert air," unless the author is discovered by some appreciative editor!

This is only excelled in point of orthography and punctuation by the following Rebel love letters which were sent from a captured post-office in Virginia, by As't. Adj. Gen. Dana to his sister in Rockford, Illinois, and which I copy *verbatim et literatim*.

REBEL LOVE LETTERS.

"PETTIGREW HOSPITAL, N. C.
MAY 27TH 64.

Dear Miss I take
the pleasure

" of writing you a few lines which will inform you that I am well as to health tho I had the sad misfortune of getting wounded on the 12th of this month tho I aint very bad

wounded I think that I will be able to go to my Regiment in
the corse of a week or less time Miss I dont feel my self
Capable of addressing a young lady that I essteem as high as
I do you tho I will do the best that I can you must excuse bad
writing and bad spelling for I am a bad hand to write or to
spell which you will see that from this letter tho I am in hops
that you wont take no insineration at bad spelt words it is
only ben about one month sense I Saw you but it seems as if
it had ben twelve month for there is no young lady that I
have ever saw in my life tho that is Saying agreat deal for
my self tho I am in hops that you will excuse my boldness
for it is my fait I gineraly think that aman on such accasion
as that ourt to be perminent in such bisseness as that for I
allways thought that aman ourt not to jest with a young Lady
on such accasion well I will close my letter for it aint verry
interresting.

WILLIAM N. HUNT.

iff you think anuff of my letter to write to me Direct to Co
H 26th Ga Regt in care of Capt H H Smith."

There's self-abasement and constancy for you, for though
he seems to have found his "fait," confesses that his "letter
aint verry interresting," but believes that "aman ourt to be
perminent in such bisseness, and not jest with a young Lady
on such acasion." Here is another equally racy:

" CAMP GREGG, VA. February 16, 1863

Dear Miss Kitty I feel ashamed of attempting to write to
you after refusing to answer your letters, but I hope that you
will not think any thing of it as you may well know I have
seen a great deal of trouble this last year. I suppose you
heard of the death of my two brothers, and I have been very
unwell the greatest part of last year, but I humbly ask par-

don for not writing, and hope that you will not refuse to grant it to one that esteems you as high as I do. I hope that the time will come soon ; when I shall return home again that I may once more see your smiling face and hear your winning voice again, oh, that I was once more a free man that I could go and come when I pleased, I hope that the war will soon come to a close for I am tired of being a soldier, I tell you that I have seen a hard time since I left home, I have marched night and day through rain and snow not only that but I have suffered hunger and thirst. My dear friend I hope that I may get home safe again. I have only written to you one time since, I have received three letters from you, in which you answered mine very satisfactory, but owing to the misfortunes that befel me I was not able to fulfil my expectations and not only that after I considered over the matter I thought best to wait until after the war, I want to hear from you very bad, and to hear what you think of the matter, my mind have not changed in the least, and I hope that you stick to your promise. Give my best respects to your mother and all inquiring friends, and reserve a full portion for yourself. I want you to write where your post-office is, I expect that you have entirely forgotten me, but I hope not, there is only two ways that any one can get a furlough here, one is to get sick, and the other is to get married, therefore, I do not expect to get one for one reason no one wont have me, and the other is I can't get sick.

Dear Miss Kitty I would give half of this world to see you and the other half too if they belonged to me for there is nothing on the earth that I prize so high and so dear as I do you, you must excuse bad writing and mistakes.

I remain your affectionate lover,

SAMUEL D. McKLODE.

26

This letter embodies quite a history.   It seems the write
after "refusing to answer her letters," wishes her to forgive
him on account of his "losing two of his brothers," "suffering hunger and thurst," and not "being very well himself."
And then she had "answered his very satisfactory," but owing to misfortune which befel him, has not been able to fulfil
his expectations, and thinks it better to "wait till after the
war."   But he "hasn't changed his mind," oh! no, although
he wants a furlough "so bad," laments that he can neither
get sick, nor marry.   Oh, the artfulness of man!   Of course
she would have forgiven him, after telling her that he would
make her a present of one half of the world when he saw her.
Here is another.

Spotssylvania Court House, Va. May 21, '64.
Miss Amanda E. Mastin.

I seat my self this morning to an swer your kind leter
whitch was dated May the 1, and caim to hand may the 20
it fown mee in fine helth and excilent spirits whitch I hop
when this few lines Reches yoar sweet smiles tha may find
you in percession of the saim kind of belesans Miss Mandy
I hav no news worth Ritin only wor news and I guess you
her a nough of that every day I will giv you a short skitch of
our tramps since we left winter qwarters we left May the 5
and went in to the fite the 6 morning at sunris and faut until
dark and the next morning we went in at day light and faut
til 10 oclock that knight when we wor Relevd and we movd
down the lin the next day had another small fite with them
and we hev bin fitin them every day since we left awn the 12
day we had the marster fite that ever has bin faut in virginia
I never Saw such a slawter in all my life well as the fite ent
over I wont say mutch about if we air in line of batel at
spotsylvania court hows and has bin fer 5 er 6 days and I ent

abel to say how lowng we will Remain her I think it is a qeer
notion to fite them here that is if tha stay to see it I wish tha
wood fite us er les go back awn the other Side of the River
so we cowd obtain Som Rest fer this army is gaded vary
mutch but air still in goad Spirits I am in fine spirits my self
all tho I field vary much exzausted fer I hent slep as mutch
as 5 knights since this thing has commenst and I hav not had
awf my shoes ner cateridge Box since the 4 day of may and
you can guess wherther we air exzausted er not I will drop
this sub ject untill the fite is over and then if I am Spaird I
will giv you a full histry of the batel well Miss Amanda you
spoak of not gittin the letter that I sent the song ballet in I
was very sorry in deed fer I wanted you to have the balet
but had Ruther you had hav got the letter fer I think it was
the best leter I ever composd in my lif and thair foar I showd
hav Rether you had a got it I want you to excus mee for the
short letter I sent buy Mr. Morris tho I thout my corispondance was not excepted but still I thout as I had sent sent you
a sheat of paper and a stamp and I cowdent help but think if
you Recieved that letter you wold hav a answered it fer it
was Such a won as you cowd not Refused I dont think I hop
my corispondance is excepted with hy Reputation as you air
the girl I think will Suit my fancy agacly tho you hav never
consented to our ceeping up a Regular corispondance whitch
I hop you will not Refus as it is the only in Joyment that a
solger sees is when he hers from the fair Sexe whitch he
hyly dos a steam giv Jan my best lov and kind Respects and
also Mr Morris and Miss Morris the saim and as fer lov I hop
you will be abel to obtain a larg potian yoar self Miss mandy
fer fere that I worry yoar patience I will bring my few bad
Remarks to a cloas hoping to her from you every weak you
will a dres mee to Rich mond co a 8 Regment Alabama volunteers in the car of captin herd I wish you to direct mee

agacly how to back yoar letters if the way that I opposed
wont suit you I hop you will excus this bad writ letter fer I
had to write it awn my nee I still remain your friend til deth.

MR. J. W. TOMBERLINSON."

What an ungrateful creature she must be, when he " sent
her a sheat of paper and a stamp," if she had not replied to
the " best leter that he ever composd in his lif," particularly
when he tells her " it is all the in Joyment a solger sees is
when he hers from the fair Sexe whitch he hyly docs a
steam!"

CHAPTER X.

JEFFERSON, GENERAL HOSPITAL, JEFFERSONVILLE,
IND., January 3, 1865.

A happy New Year. It is pleasant to chronicle an act of disinterested benevolence. A Mr. Kisling, of Delaware Co., Ind., visited his son in this hospital, bringing with him some creature comforts. While his son was enjoying them, he heard others wishing they had friends to do the same for them. He immediately formed the resolution of seeing what could be done in that direction; and upon returning home succeeded in interesting the good people of Delaware and Henry Counties, Ind., in his project, who responded liberally. And the result was 400 *chickens*, not veterans, innumerable cakes, with pies, fresh butter, onions, apple butter, and canned fruit, in all about 3,500 pounds were contributed for the benefit of our boys. It was enjoyed all the more from its being a surprise. Three capital holiday dinners in succession, well, the invalids of this hospital do not need to sigh for home, on account of these festivals. Blessings attend the donors!

In striking contrast to such fare was that of one of our patients who came from Nashville last Saturday. He was taken prisoner at Franklin, and was with the enemy eighteen days. During that time he was an inmate of the hospital, having a wound through neck and shoulder. His fare and that of the other patients was two meals per day, two biscuits of hard tack, and one piece of meat in all, for the twenty-four hours. There were at first 287 Union prisoners in the hospital. Five physicians and eleven nurses were detailed for duty; but all except one of the former and four of the latter deserted. Of course the patients were the sufferers by this desertion, and out of the original number but 163 survived. Not all of this number were in one hospital, but of those in the one in which was my informant, fifteen died in twenty-four hours.

He says had it not have been for the kindness of a young lady by the name of Fannie Courtney and her mother, in bringing in baskets of provisions, he believes some would have starved.

Over 200 patients came from Nashville to this hospital last Saturday, out of which our ward received more than one for every bed. We have fifty-nine beds for patients, and we had seventy-one on the morning report for the new year. Mattresses are put on the floor at night. About the same number came into the hospital on Sunday, and half as many to-day. In a few days there will be a transfer from this to some other hospital farther north, to make room for others from the front. Immediately after a large transfer to a hospital the greatest mortality occurs. They are sometimes brought in a wretched condition. Some have the balls remaining in the wounds. One here has not slept for three nights from that cause. Another came whose wound has not been dressed for thirty-six hours, and as a consequence he has gangrene. Not long since two men showed me their shirts which had been worn without change between twenty and thirty days. And there was no help for it until the next week, for at that time we had but twenty-five clean shirts for seventy-seven men. The hospital is overcrowded. It was only intended for 2,000, but we have had upwards of 2,600. And orders have been sent Maj. Goldsmith to enlarge it so that it may contain 5,000.

It is evening; I am seated in my ward by one of the four mammoth stoves, which are ranged at equal distances through the length of the same. Groups are gathered around each

stove. Some are chatting on army experience, some discussing politics, some dozing in chairs, perhaps a third of the
whole in bed, two playing chess, one singing low to himself, as
if to pass away time, and last, but by no means least in her
own estimation, is one who dips occasionally into the inkstand
which is upon the same table as the chess-board, and is occu-
pying herself in "telling tales out of school." Here is a
short one.

A German boy sitting just behind me on the bed who has
an "interesting" arm, has just been telling me the following:
"One day when our regiment was down in Georgia, a party
from our company were out on a foraging expedition, and
came to a house where were a woman and her two daughters.
As we rode up the mother held up her hands in blank astonishment.

" ' Why, youans beant Yankee soldiers, be ye ?'

" ' Yes, we are !' was the emphatic reply.

" ' Why, youans looks like weuns do !—only I don't know
but youans looks better'n weuns."

" We tried to make her tell," says my informant, " what
weuns were expected to look like, but without success."

JANUARY 6.

Day before yesterday I was very busy with the little er-
rands which seem almost nothing in the telling, but yet effect
very much the comfort of the invalids, when an order came
for " seven men to go from each ward to Jefferson Barracks,
Missouri, who were from the States of Missouri, Minnesota,
Iowa or Wisconsin—the number to be made up, if necessary,
from Illinois." I was permitted by the surgeon to go through
the ward and see who wished to be transferred nearer home.
Sometimes, so little time is given after such an order is issued,
that with the pressing duties of a surgeon it is impossible to

take time for the selection, and some are sent farther away from their homes, while others are retained who would gladly go.

Then came the furlough of eight of the patients, several of which I had looked for for some three weeks, almost as anxiously as the recipients. These same furloughs may be the means of saving three or four lives. This chronic camp disease, or the scurvy, is best cured at home, if at all, and if the patient is not to recover, he is certainly better at home in most instances. It was fortunately not very cold weather, as only one had an overcoat to wear away. He very fortunately had his descriptive list and drew one. The others had nothing except blouses over flannel shirts.

One man, an Indian, from Michigan, had lost all his baggage. What was he to do without money or a descriptive roll, not having been here long enough even to get two months' pay? Fortunately I had one shirt remaining from the January stock of flannels. Not one was left in the Sanitary stores or about the hospital, except some two or three which some ladies had sent from home. I had already borrowed three such pairs of socks for those who were going to the front. The Sanitary Agent had told me she had but two or three pairs of socks to give out only in extreme cases and a few flannel drawers. This, I thought an extreme case, and taking the last flannel shirt in my possession and going to the lady I received the other two articles, and hastening to the ward gave them to the worthy representative—not of the copper-headed but of the copper-colored race. The ward-master secured other clothing for him, so that he was as comfortable as was necessary.

Another young Indian, who could speak but a few words of English, received a letter to show to " whom it may concern," asking assistance if needed ; while by sign I think he was made to understand its purport. I had been to head-quarters and obtained orders for an ambulance to take the men to the depot for the evening train of 8.40. Then left the ward after they had received their baggage and rations. Learned this morning that the ambulance did not come for them, and that all, except one crippled boy, had walked a mile through the mud to the depot. Thus are men nerved with additional strength when the stimulus is home and loved ones.

JANUARY 15.

Wonder how many people at the North think we are living on champagne and canned fruits at Uncle Sam's expense. Wish such could see our table. Please imagine, dear friends, your humble servant as sitting down to a long table with some eighteen others—not tables but ladies—and viewing three plates of bread, three bowls of gravy, ditto of apple-sauce arranged at equal distances, and that each has the exquisite pleasure of chewing for a reasonable length of time a piece of tough meat which is strongly suspected of having once been the person of a mule, and of drinking a mug of coffee minus the milk—and oh! worse than all the rest, the table is minus the butter. These two last are regretted the most. I wish somebody would make a raid and capture a dairy—milk-maid and all! Won't some good Northern body be so mag-nanimous as to send me a little pat of butter and a cup of milk?

The truth is, I have encountered perils by land and sea,—thrice being obliged to do my own washing save once, because the laundress had married a husband and could not come. Once in my life did I have the audacity to pay special atten-tion to a young corporal from Massachusetts by accompanying him to church one Sabbath evening, and came very near

27

being discharged for the same. Shall never dare to repeat the heinous offence. Special attentions not allowed among Uncle Sam's nephews and nieces. It is my opinion that said corporal is not over fifteen years younger than myself, still there's no knowing what might have come of it. Ah, me! what a sacrifice am I making for the good of my country.

"Lady nurses in the hospital," says a late order which was sent to each ward, "are expected to be in their wards each meal time to receive the special and extra diets and deal them out; take charge of all liquors used in the wards, and do anything else required by their surgeons."

This, by the by, is just what we've been doing all the while, especially taking care of all the liquors we can obtain, in addition to which some of us have done whatsoever our hands, heads, or hearts might find to do, whether ordered by a surgeon, or not.

But to return to the fruitful subject of our own diet. Perhaps I have colored the matter a little too highly, and to be just, will mention that for some three consecutive mornings we have had the exquisite felicity of inhaling the fumes which arose from buckwheat cakes, just after arrival of that kind of flour to the hospital. Said fumes issued from the steward's mess just across the hall. It was so grateful to the olfactory nerves, we thought of sending a deputation to wait upon their dignities and humbly request a continuance of the same for our benefit. Once, since our sojourn here, somebody has has had a remarkably severe fit of benevolence, not fatal, however; and the consequence was we had hot rolls for three successive mornings. If only somebody would send one of those nice little needle-books, or comfort bags, with an affectionate letter in the interior thereof from some "nice young lady at the North," and I could present the same with one of my most winning smiles, and sweetest tones, mayhap we might have hot rolls for three more mornings, or what would be still better,

> Have butter on our daily bread,
> And milk within our coffee.

This poetry is perfectly original ; please don't anybody claim it, on their peril.

Again, I confess I may have colored the matter of the diet too highly, for there has been an overplus of butter in the wards since it has disappeared from our table, and conse-. quently some of us have seen fit to capture a piece ; therefore, there may be seen by the side of some plates a little pat of butter done up in a rag. Happy possessor of the rag with butter in it, even if it ranks higher than any general in the field !

JANUARY 17.

Am feeling grieved and sad this morning. The chief nurse is to-day sent to the front. What a pity he could not with-stand the temptation which sparkles in the wine cup. A more capable, prompt and cheerful nurse is seldom or never found I wrote a few lines to him, and received a reply. Will copy both.

"WARD 1, January 17.

" Mr. ——— :

" MY DEAR FRIEND :—Do you know some of us are feeling sad and grieved this pleasant morning, and do you know the reason ? I snatch a moment to write you a few hasty lines, even at the risk of not knowing whether or not they will be kindly received. In addition to the feeling which prevails in the ward, of regret that one so prompt, cheerful and capable in a sick room, is going to leave us, is another, which, even at

the risk of wounding your feelings, I must express, as I have
a good object in view. It is, that one so capable of making a
noble specimen of manhood and of doing so much good in the
world, is liable to be bound in chains stronger than ever ty-
rant bound a slave. Do you know, my friend, that 'he who
ruleth his own spirit is greater than he who taketh a city?'
Cannot you battle bravely against this one temptation, and
prove yourself the conqueror? I am a member of the
Order of Good Templars, and I wish very much you were
situated so that you could unite with that, for it might be a
great help to you.

" I have used my influence to keep you here, and would go
myself to the executive officers and intercede, but that I know
there is sensitiveness upon the subject of 'interference of the
lady nurses.' It is universally regretted in the ward that you
are to leave, and some have interceded for you. Dr. C. also
regrets the fact, but he has several times come between Dr.
M. and the men in his ward, and once before on your account,
and farther intercession would compromise his own position
and influence.

"If this is received in the same spirit of kindness in which
it is written, I shall be glad to know. I shall be glad to hear
from you, from any place where you may be. But the bugle
sounds for dinner and I must close.

"Your sincere friend,

———   ———"

"GUARD HOUSE, January 17.

" Miss P———:

" MY DEAR FRIEND:—I received your kind and welcome
note a few moments since, and am very thankful for your advice. I will make a promise but I do not know as I can keep
it, but will try, and if I can govern myself shall ever be

thankful for your advice. You know that the army is a hard place for young men, and we are always tempted to use this poisonous stuff.

It took me down when I read that letter, and made me ashamed of myself to think that after I have done as I have, you would write to me and give me such good advice. And I have resolved not to be ruled by that one temptation, but to battle against it and attain the mastery over it. I will keep that letter of yours, and when tempted to break my resolution will take that letter and read it.

I could have come in the ward this morning, but did not wish to, as it would make me feel so bad. I shall have to bid you good bye from here. I will write when I get to a stopping place, and shall always be glad to hear from you.

" From your friend,

_____ _____ "

JANUARY 22.

Nearly 1,000 patients have been added to the hospital within the last ten days. The " R. C. Wood " brought up 700 and left 100 at New Albany. Two days after, the little hospital boat " Jennie Hopkins," brought 269 more. From this number should be taken, however, fourteen, who died on the passage, nine on the barge, and five on the small boat. As many more died within twenty-four hours after their arrival.

" War is cruel, and cannot be refined," was the defensive shot fired by Sherman at Atlanta. Still it seems a pity that men should be sent out from Nashville hospitals, in a dying condition, to make way for rebel prisoners. Why could not some private mansion be used for that purpose, whose owners are known to have taken the oath merely to save their property? At the door of one such residence during my stay in that city, one of the young ladies was heard to say indig-

28

nantly, " Well, *we* shall have to leave here before long, that's sure, for I see no signs of these *Yankees* leaving." Or very good use might be made of that of another, who confessed to my informant that she was "*obliged* to take the oath on account of her property, but that if her son enlisted in the Federal army to fight against his friends, she would herself take his life with a revolver."

I did not visit the large boat myself, but a lady who has been connected with hospitals over two years, a good part of the time at Memphis, and not of the sensational stamp, says that she never before saw such scenes of suffering.

In company with this lady and one other, I visited the " Jennie Hopkins," the next morning after her arrival. All had been taken from the lower cabin the night before, and for some yards from the door of the corridor to the place where the ambulance stopped, the snow was red with the blood that had dripped from the wounds of the patients. As we neared the boat they were bringing off patients on stretchers to the ambulances, while others were walking. Among those on litters, was one little lump of humanity so small, enveloped in blankets, as to make me doubt whether there was anything but a blanket. Probably a wee bit of a drummer boy, I thought ; and it might have been the same little fellow of fifteen years, who I have since learned was taken into one of the wards, without any clothing except blankets, and perfectly benumbed, and who died the next day. Among those who walked, I saw two middle-aged men, with their arms supporting each other, looking so pale and emaciated as to make me wonder whether with such weak and uncertain steps they would ever reach the hospital.

Upon entering the boat we saw nothing particularly striking to those who are accustomed to hospital scenes. They were about lifting another man on a stretcher, when a surgeon told them they must wait awhile as the ambulances were full and gone.

" What, only three ambulances to such a big hospital as this!" said one patient. " Oh, dear! seems like we'd never get off," said another.

" Oh, carry me home, oh, carry me home,"

sang out another, whose healthy, jovial face enveloped in a turned-up overcoat collar, seemed in striking contrast to the average of the company. But a pair of crutches lying upon the bed beside him, gave proof of his claim to make one of this number.

" We'll have chance to stay here to dinner, yet, to say nothing of a luncheon," he mischievously continued. Patients on these boats complained bitterly of their fare.

" The most ravenous set I ever saw," said the man, having charge of the full diet in our ward, two days afterward; " they must have nearly starved them."

They commenced dressing one man, evidently very low, drawing on his clothes very slowly. Then they paused, and three or four gathered around. The steward left him, came past us, unlocked a door, and taking a glass drew some brandy from a keg. " Is he dying?" was inquired in a low tone. " No, but very low," was the reply. The liquor was given, and he revived so as to be taken to the hospital, but very likely was included in that number who died within twenty-four hours. I profess to be a Good Templar, but I sometimes thank the Lord fervently for good brandy. To die in a pleasant ward, on a clean bed, and with every needed comfort, and a letter to the loved ones at home, with even this poor comfort for them was better than to die there.

We passed around and spoke to several; and when I saw one boy of about eighteen, from a distance, I said to myself,

"Surely that boy is able to go to the front;" but upon approaching, saw what was the matter.

"Ah," was the query, "how long since that foot of yours left you? You're *looking* well enough to go to the front."

"I *am*," said he, with a genial smile on his rosy face. "It's only about three weeks since my foot was taken off, but I havn't been sick a day. I am *well* enough."

And such is the difference. Blessed are those who go into the army with pure blood, sound constitutions, and a habit of looking on the bright side.

One middle-aged man, whose countenance as I read it, told of sterling worth and stability of purpose, was lying in bed, and with glistening eyes, he told me how greatly he had suffered, as he showed me the heavy, jagged minie ball which had ploughed through the bones of his ancle.

"It was at the second charge on the 15th, and of the 16th army corps, at Nashville. I had fired my piece, and had dropped on my knee to reload, for I was but a few yards from the rebel breastworks, when the ball struck me. I was taken to hospital No 1; but on the night of the 20th, our men were all taken out of the hospital to make room for 1,500 rebel prisoners. There was a cold sleet at the time. I took cold by being taken to a tent where I suffered dreadfully. I can never tell how much I endured there for several days."

Another man who lay near, showed me a rough three-sided piece of iron, weighing about a quarter of a pound, which he told me was "the piece of shell which laid him up." He corroborated the statement of the other, about the removal from the hospital. Several in my ward have told me the same thing, who were put on the next train and brought here. "It was understood," say they, "that we were taken out to make way for 1,500 wounded rebel prisoners." And it is the testimony of surgeons here, that the greater portion of those who came on those boats, "were not fit subjects for transfer."

If this taking of sick and wounded men from warm quarters, and sending them off to suffer and die, that rebels may be made comfortable, is a military necessity, I " don't see it," as the soldiers say. "There were the greatest number of recent amputations, on the large boat, I ever saw together," said the lady formerly alluded to, who visited it.

" Fresh amputations, where arms had been taken off close to the shoulder, were sent off on transfers," says a young theological student, at present a patient in my ward. Here is more evidence. I was at the gangrene tents and barracks soon after, and one in the latter place said, " Oh, my wound was looking so nice when I left Nashville, but there were 150 of us, whose wounds were all washed with one sponge, and none were dressed but once a day, and I think as many as fifty took gangrene. There were five who did in this room." I inquired of those, and found it their testimony also, as well as of others, still, who were at the tents. One told me he had gangrene when he left Nashville.

As far as regards my own ward, one was brought here, with gangrene in his wound, which was contracted on the boat. He was brought in at eleven o'clock one day, and died at eleven the next night. I was standing by his bed some two hours after his arrival, taking the name of "nearest relative " to write to for him, when he casually mentioned that he " did have a brother in this place some time ago, but hadn't heard from him in a month, and presumed he had been transferred, or, perhaps," said he faintly, "he may be dead, he wrote he was pretty sick. He was in ward 10, and, I think, Hospital No. 8. I had it in a letter, but I threw it away, for I never expected to be here."

" We have no Hospital No. 8, in the place," I said, " but are you sure about the ward ?"

" Yes," he was sure the ward was No. 10.

"Then if your brother is in this hospital, I can find him," but not to excite him too much, I added that very likely, as several transfers had taken place within that time, he had been sent away. I started out immediately, went to No. 10, called for the ward-master, and found that the man I was seeking, was well, and had been detailed for duty in the bakery. I started again on the covered corridor, just now such a convenient protection against the snow-storm, passed my own ward, and went to the bakery. I opened the door and saw some three men within, who wore the badge of their present employment, in the shape of flour on their clothing.

"Is Mr. Moses C —— here?" was the inquiry.

"Yes, that is my name," said a young man stepping forward. His manner and tone were seriously expectant, why, was soon understood.

"You have a brother here," I said.

"What, not a *wounded* brother?" he said excitedly.

"Yes," I said; "he came on the boat to-day."

"Why, I have a brother here now, just from home, who had started for Nashville to get his body, for they had heard that he was dead. Where is he? What ward is he in?" he continued, moving towards the door.

"I will show you," was the reply, as we started out; "but your brother is quite low, and it will be necessary to calm yourself, so as to excite him as little as possible."

He promised to do so, and I made him wait at the upper part of the ward, while I went to the lower, and broke the news to the dying man. It was a blessing for me to see the kiss given and received by the dying lips, and for the three brothers it was a blessed privilege to meet and commune that day and the next, before death hid one of them in the grave.

Since I have been writing this, while sitting at my little stand, near the door of the ward, a man came quietly in, who paused, waiting. I looked up and saw Mr. King, the one mentioned under date of December 21st, who had been home on a furlough to get his starving family out of rebeldom. After the usual salutations, I learned the sequel to that before noted, and the burden of the story condensed, ran thus:

" My wife got the five dollars you sent, so she and the children did very well till I got home. The guerrillas came into town again, last Friday was a week, but I was six miles from home at the time. They came and robbed the stores and were away again. They whipped in within a mile of a Federal regiment. They knew when to come mighty well; when the Unionists ain't to home, somebody tells 'em. The Union folks is all leavin' the place, anyhow, and its next to impossible to sell anything. I sold my cow for forty dollars, and two or three other things that didn't 'mount to much, an' it cost me fifty dollars to get my family here. Then I've had to pay 'bout six dollars for tub, wash-board and some other things we couldn't keep house without, and I've but a little left. Do you know if I'll be able to get any pay this time ? "

" Were you here the first of November ?"

" No, I came the ninth."

" And your descriptive roll is with your regiment ?"

" Yes."

" Then, I'm sorry to say, you'll get no pay this time."

Then he hung his head, and said, meditatively to himself rather than to me, " I don't know how I'll get along then. I got here in town," he continued, "about a week ago, but my furlough wasn't up yet, and I stayed to the Refugee House. I've been a-takin' keer o' them thar, and givin' 'em medicine, for thar ain't scarcely one thar what can read any sort o' handwritin', an I brought a recommend from them, axin' Dr. G. here, if I can't have a permanent detail an' stay thar. My furlough's just out to-day, so I come up to report an' ax for the detail."

So much for his story, one among many.

At the present date, I have but two or three very sick men in the ward, and the one who is probably the worst tells me that he is " right smart better, this morning." Still one very bad feature, as connected with the institution, has developed itself within the past forty-eight hours, in the shape of some seven cases of small-pox. They are being removed to the proper hospital at Louisville.

SATURDAY, 28.

Charles Stearns, Co. F, 177th Ohio, was broken out this morning with what seemed to me measles. Dr. D. soon came in, who is taking the place of the ward surgeon for some twenty days, he having gone home on furlough. As soon a I saw the eruption I felt nervously apprehensive of what may be the result, for I remembered the sixty cases of measles in Nashville, which were sent to the small-pox hospital. I told the physician of this, and my belief that the man had measles. But he thought otherwise, and the patient was taken from his warm bed close by the fire, and carried to a tent where was a small-pox patient. I hope the surgeon was wise in removing the patient, but I fear the worst. Vaccination is since being rapidly performed.

Sometimes they make queer mistakes in the reports. One man was detailed for light duty who was in the dead-house. Others have been ordered to the front who were on a death-bed, and others are reported dead, who are well.

The usual punishment for almost any misdemeanor is the guard-house, and then the front. Though sometimes, it is the carrying of a large stick of wood on the shoulder, or a board strapped on the back with the word " Drunkard," back and forth on a certain walk, so many hours per day.

One man from my ward, " the fancy fellow," as the boys

called him, because he dressed so well, refused one very cold Sunday morning to help carry coal for our fires, and was sent to the guard-house and then to his regiment. Another stole a pocket-book and met with the same fate. At the time of the New Years' dinner, it was stored over night in the dining-room, of ward 12. The boxes were broken open that eve, and the goodies were being devoured and secreted that night and next morning. Miss H. and somebody else, chanced to learn the fact, and it was reported to the executive officers. It was found to be two kitchen boys, and one patient who was having extra diet carried to him. Chickens and butter were found secreted in their overcoats.

These were promptly sent to the front. One man was sent to the guard-house for spitting on the corridor. On Christ-mas, as a man was going into the large dining-hall to dinner, who had just arrived from Nashville, he said aloud that they were "going to have a feast then, and live on crumbs and scraps, or starve for a week, to make up." Dr. Mathewson, the executive officer, overheard the grumbler, and he took bread and water in the guard-house instead of a Christmas din-ner.

Two men have just been telling me of a little story of one of those wise refugees. It was near Kenesaw Mt., and before the battle there. The regiment had a large tin steam oven in their possession. This was mounted upon a wagon, and as it passed a house, a woman with open-mouthed wonder in-quired,

"What that thar thing was."

"That—oh, that is one of Gen. Sherman's flanking ma-chines," said one with the most impurturbable gravity.

After the battle of Kenesaw Mt. they passed that way again, and one who had known of this took occasion to speak to the woman of the result of the battle, and she replied:

"Wall, its no wonder weuns got whipped for Mr. Hood

haint got none o' them flankin' machines what Mr. Sherman has. I don't b'lieve Mr. Hood could get one o' them are things in all this ere country !"

The hearer, also, " thought he couldn't."

TUESDAY 31.

" Charley," the small-pox patient, was sent to the small-pox hospital at Louisville yesterday. I still think it measles.

Saw a letter last eve, which had originally been directed to Cumberland Hospital, Nashville. That was erased, and in red ink appeared the words, " transferred to Louisville, Kentucky." There it had been marked " Clay Hospital, not here," " Crittendon Hospital, not here," Then " Jefferson Hospital, Ward 7." That was crossed out, and the words written " Tent C. Gangrene." To-day that letter was carried to the place, by Miss French, and as she read the name aloud one man exclaimed, joyfully, " That's *my* name," then as he received it, he said, with streaming eyes, " God bless it ! "

FEBRUARY 5.

Yesterday morning a stalwart, healthy man, stood guard on the corridor at the carriage crossing, who at night was lying in the dead-house, the victim of drunkenness.

Dr. C. came back yesterday from his furlough. Patients all rejoiced to see him. He says " Charley," the ward-master, and I both look as if we had been sick. We have had a great deal to do and are quite worn out.

First sermon in the new chapel to-day. It is not plastered and was quite cold. Everybody sang a tune of their own. The building of this chapel is Chaplain O's special care and great anxiety. And though he has several hundred volumes of books which have been donated for the use of the patients of this hospital, yet he keeps them all closely boxed up, until the reading-room in the chapel is finished, which may not be done before the hospital is broken up. It is such a pity when there are so many hundred men passing through here who need them.

Have just learned that my diagnosis of the so-called "small-pox patient," was correct. It was a bitter cold day, the river was frozen, and the ferry-boats not running, and he was taken via New Albany to the small-pox hospital. He was carried the twelve miles in an ambulance in his bed, and without being dressed. He took cold, and was kept only thirty-six hours at that hospital, when it was decided that he had measles and he was sent to the measles hospital. Last night when his brother visited him, he was not conscious. He cannot live. He has fallen a victim to what?

February 7. He died yesterday.

FEBRUARY 11.

Two men started home on a furlough to-day. One of them W. C. Stewart, Co. I. 7th Ill. Cav., it seems is of a family of heroes. His father was in the revolutionary war, his oldest son was in the Mexican, was wounded and exempt in this. But the father says he did not say to his other three sons "go" but "come," and went with them. One was killed at the battle of Corinth. He tells me that when home on a veteran furlough, the copperheads of the place had determined the soldiers should not vote. "But," said he, "we gave them to understand we would fight for the right if needful, and some eight of us armed ourselves and marched to the polls, and every one voted." He says also that one of his sons was offered $1,000, to go as substitute, "but" said he, as the determination of the patriot blazed in his eyes, though over sixty years of age and unable to stand without leaning on both crutches, "If he had been *bought* to stand up as a mark for rebel bullets for another man, after fighting them so bravely as he had, I'd have been almost tempted to have shot him myself. He re-enlisted to fight for himself. This furlough has been a long time coming, but if I get home in time to see the copperheads squirm when the draft comes off, it 'll do."

Miss McNett says a humorous patient in her ward, who has eaten almost nothing for a day or two, upon her asking what he would have said, " Oh, almost any thing, if it has a woman's finger in it."

One middle-aged man at the gangrene ward told me last Sunday that it " *did* do the men *so much* good, to have a woman come, if she didn't say more 'n one word, it revived 'em so ;" and he earnestly appealed to the ward-master if it was not the case, and who agreed with him.

In contrast to this was the assurance of a surgeon in one of the wards to the lady, Mrs. C., that the lady nurses here, were regarded by the generality of the surgeons as " permitted nuisances." Nevertheless I am strongly of the opinion that if either of the surgeons should be *really* sick, they would be very glad to get " something with a woman's finger in it," even though not frank enough to own it. And some of us at least came only to take care of the sick, and care much more for their approval, than for any slights which can be given by others.

Mrs. R. is joyfully elated this morning, for she tells us earnestly that her " men *all complain of being better*. On the contrary one of mine informs me that he has " got a big misery in his breast," another that he is " a *heap better* than yesterday," and another that he's *right smart better*, though *powerful weak yet*, thank you madam."

Don't know whether he has me to thank for all of that, or not.

An order has just been received from the surgeon-general, to the effect that no lady nurses shall be kept in hospitals, except soldier's wives and widows.

Dr. C., but especially Mr. Bayne, say they shall have to hunt me up a soldier, and the latter inquires seriously, and with a very fatherly air, if "red whiskers will be a serious objection." I tell him it will "not be an insuperable objection, as I expect to make sacrifices for the good of my country."

FEBRUARY 17.

Yesterday was an April day in my calendar. The showers came when Mr. B., one of nature's noblemen, a gentleman and a scholar, albeit our "kitchen man," and also honest, warm-hearted and cheerful Peter, chief wound-dresser, were ordered to their own State, N. J. It was a matter of regret all around, to themselves and everybody else. It seems a pity that such responsible positions as chief nurse and wound-dresser, those who by long experience know their duties and have the confidence of the invalids, should be made so light of. These same men, after reporting in their own State, may remain in some hospital for months without having any part in the work for which they are so well fitted. But the order came from a superior officer, ordering "all New Jersey men who were able to bear transportation to report at Washington." So much for the showers.

The sunshine was poured on amid the showers by the arrival of the Rev. D. P. Livermore, from Chicago, with *seven* boxes and a barrel of sanitary stores, for Mrs. Colton and myself. These contained a nice supply of flannels, dressing-gowns, rags for wounds and dried and canned fruit. Wasn't I overjoyed? could hardly sleep last night.

FEBRUARY 23.

Have been urged, on account of my failing health, to ac-
30

company a friend to her home for a rest of ten days. Had decided to do so, but the ward-master is taken sick, threatened with fever, and one other poor boy is running down so fast I feel that I must stay if possible. Will try to get well here, and attend to them also. Sick the last twenty-four hours.

Patients in this hospital do not think much of other hospitals in comparison. As one evidence among many, will give extracts from two letters just received, one by myself another by a patient, from our lamented Mr. B. He writes thus graphically:

"WARD 3, NEWARK, N. J., FEBRUARY 19, 1865.

My Dear Miss P.—We duly reached this delectable dumping-ground, after fifty-six hours of almost incessant motion. The establishment consists of an ancient tannery, located picturesquely among lumber-yards, railroads, debris of all kinds, and the Passaic river. The result fulfills my premonitions, I can only pray that my stay may be short.

I have no doubt that the loss of my valuable endeavors at the Jefferson, in the artistic arrangement of bread and molasses, has proved irreparable. But I am consoled by the reflection that this is not the first blunder, evincing lack of statesmanship, made by Lincoln's administration during this war. What the consequences are likely to be, it would not be safe to predict.

I think Dr. Mathewson must be a miserable man, coming in as he does, and breaking up the civil, political and social relations of men and women, as good as himself. Don't you think so too? It seems impossible that he should sleep well o'nights. Ergo, he must be splenetic and dyspeptic in the morning. Ergo, he must be very unhappy. I believe too he has black whiskers, and I have read in that highly exciting, historical romance, entitled "The Bloody Shoe String, or the

Murdered Milkmaid," that pirates and assassins always have black whiskers. I leave the reader to draw his own inference. Very different is Dr. Chapman's unhappiness. It is of that godly, (or goodly-something) sort, which Paul tells about, that one need not be sorry for."

" Upon our arrival, we were immediately regaled with cold tea, stale bread and strong butter. We were then shown up to a loft in the building, and given beds filled with straw, with but one sheet, ditto blanket, but as there were no open windows, or ventillators, and we were very tired we snored away very comfortably till morning. At breakfast we were regaled with rye coffee, and stale bread and the aforesaid strong butter; after which about twenty doctors, headed by a small man with a sword, marched through the ward, the little man calling out " salute, salute," as he traveled along.

The little doctor was the surgeon-in-charge. This place is just what I knew it was. If a man blows his nose too loud, he goes to the guard-house, and there is $5.00 reward for telling who spits on the floor. I don't know what they do with them. Very likely they are drowned in the river close by, for I don't see what else the river was put there for. Last week, I am told, two men became so disgusted with the place, that one shot himself and the other hung himself, and others are thinking seriously of the same thing.

Depend upon it New Jersey is a great country, if it was only white-washed and fenced in.

This morning I asked the doctor for a pass for forty-eight hours. But he assured me that I might run away, and never come back. Then I asked for a pass to go out and see the town, but he could not attend to me then. Peter, however is out, and B. S. who you remember went from Ward 1 on a furlough, has not reported at all, and is going about town.

The nurses and ward-master here are citizens employed by

Government at $25 per month, and a filthy, saucy crowd they are. The wardmaster in my ward is an Irishman who cannot read or write correctly. Every body must do his own washing, or hire it done, and must find all his own clothes. They have neither slippers, nor gowns, and every man tumbles into his straw-bed, when he arrives, with just what he has on his back. We have no women in the wards, and I don't see as they have any light diet. The men that are not able to get down stairs have the same food brought to them as is given the convalescents. The ward doctor sits in his office, and the men that are able to walk must go to him for their quinine.

Take it all in all, this is the greatest institution I have ever visited. It should by all means have the leather medal. This is the stripe of the United States army hospitals, to the Eastward, within ten miles of New York. Don't you wish you was here? It is so nice."

Signed,

———— ————."

MARCH 2.

Willie B. says it is necessary for a man to get so that he weighs 180 pounds, before he can be admitted to the invalid corps. He is an Alabamian, and has been telling me of his escape from home. He says:

"We had hid, and laid out in the woods for ten or twelve months, and were tired of it. There were nine boys of us. We travelled fifteen miles the first night, and in the morning, soon after the last of the boys had joined us, we reached Sand Mt., and after a little while we heard a horn in the valley, and we thought in a minute what's up. And sure enough, we were right, and the bloodhounds and the hunters came on after us. The dogs had a strap of leather round their necks and an iron rod to each couple, parting them about a foot and a half. Then we started in earnest, and one of the boys said "let us set fire to the woods." Then we made for that side of the mountain where the woods were, and set them on fire and then waited till the dogs lost our trail in the ashes and set out on the side from which the wind came. We travelled five nights, hiding by day, and reached the Union lines at Bridgeport, where we all enlisted."

Another death of one of our members occurred some time since, which I neglected to note in its proper place. He was a German, Valentine Rowe, of the 72d Ohio. He had been a great sufferer, had been twice out to the gangrene tents and suffered greatly from burning and hemorrhage. He was a long time dying, did not know it, but "wondered when that pain would ever get stopped in his chest."

Warfel was just telling me of the narrow escape of one of our nurses last night. He was on duty as guard and nurse in the ward, but had lain down and fallen asleep. When the officer of the day came in, whose duty it is to visit every sick ward at midnight, and who chanced to be Dr. D. who attended here in the absence of the ward surgeon on furlough, he called out " nurse, nurse," then added, " I shall have to report him but I hate to do it ; perhaps," he added, " he has gone out after coal."

So he passed down through the ward to the ward-master's room and closed the door behind him. Then one of the patients, a paralytic, having a little bag of salt which he had to use with eggs, threw it on the nurse and waked him. He suspected something of the truth and started up, though more asleep than awake. " Run," said one " catch up the coal-bucket and run out." But he had just taken it up when hearing the back door open, upon the hurried advice of another he commenced the vigorous filling of a stove. The docter came up and said good-naturedly :

"Oh, you were out for coal, wasn't you ?"

I am sorry to record here, that very unwelcome fact, that one soldier at least was known to perpetrate an untruth.

The patients showed by their action in the matter that he was considered at least " worth his salt."

SUNDAY 5.

Have been reading to C. T. Bryant, from "Stumbling Blocks," by Gail Hamilton. This patient has lain on his bed over two months in this ward, from a wound received at Nashville.

Also attended the baptism of a young man by the name of Ray, from Niles, Michigan. His sister is with him. She has got discharge papers just made out for him a few hours since, and he was so anxious to get home to die. But his death has been hourly expected the last twenty-four hours.

He was sick some months, about a year since and received a discharge furlough. He was without money, and a lawyer at home offered to loan him some, and take his papers and draw the pay for him. The papers were mislaid and lost. Then he was taken for a deserter, and carried in irons to Louisville. There he was released, as a file of the furlough appeared on the books, but instead of being allowed his discharge as had been promised, he was sent to his regiment.

The disease was checked only, and it has brought him here to die.

MARCH 10.

River very high yesterday, up to the second story of some houses in Louisville, but this hospital is on Mount Arrarat.

It froze last night, and is "right cold" to-day, as Illinoisans express it. Old winter is giving us a parting grip.

Mrs. C. has been telling me one or two incidents which I will note down ; she has lived in Missouri and Louisiana. In crossing the plains, as they stopped at a place they inquired the news.

" Well," said one in a whining voice, who had said he was from that famous place, of " Hooppole township, Rosey County, Indiana," " Lee has whipped the Federals all to pieces."

" You lie, sir," said Mrs. C. quite emphatically, and besides you're a copperhead and rebel sympathizer."

" Oh, you're too hard on the man," said a gentleman of her own party, " we don't know but the report is correct, or if not, he may have told it as he heard it."

" I say he is a copperhead," she affirmed, " for when you hear a man say ' Lee has whipped the Feds all to pieces,' and say it as if he enjoyed it—and besides he was looking down and digging his toes into the ground when he said it—its safe to pronounce him one. And," she continued, " I'll wager what money I have against a penny, that if we ask those people who are coming what the news is, we shall get a different report, for just after a rebel defeat, you'll always hear copperheads relate the dispatches which come through their grape-vine telegraph."

A party here rode up, and selecting one who wore the garb of a Union soldier, which contrasted with the butternut clothes of the sympathizer, she said :

" Sir, I recognize you as a United States soldier by your dress, will you be kind enough to tell us the latest news ?"
" Madam," was the reply, " the very latest news as I understand it, is that Grant has got Lee just where he wants him."

I note this, not merely as an incident to remember, but more as a reminiscence for myself and as characteristic of the woman. She is one of our eloquent and praying Christians, but of a strong and easily kindled temperament. She is of Southern blood and her father was a slaveholder, but no stronger abolitionist can be found than herself.

She tells me that a Dr. Dods, of Clark County, Missouri, was visited one day by rebels. Doctor and his wife had seen them coming, and she had told him to hide in the corn-field, and supposed he had done so, when instead, he had gone to the cellar. His wife, upon their asking to be shown over the house, manifested the greatest willingness and lighted them to the cellar, and went round with them telling what was in this barrel and what in that. Her husband was lying behind the one containing vinegar." " I believe this is cider," said one of the men laying his hand on the barrel.

" No, it is vinegar," said the wife, and both passed on, she supposing it policy to keep them there as long as possible, but the doctor was not discovered. The lady is first cousin to J. C. Breckenridge, and the doctor the same to Mrs. Lincoln.

Mrs. Rumsey says in making some artificial flowers for her ward, she remarked to some of the patients who were near, that such a flower was " the emblem of innocence and purity." " Oh, fie !" said one, " innocence and purity are about played out in the army."

This is about equal to the remark of another, in speaking of the practice by Chaplain Fitch of holding service in the wards and praying with the sick men, he said : " The Chaplain knows what he's about, he's just playing off."

The comic or ludicrous is often mixed up with the serious, here as elsewhere. Mrs. R. says that the other day as herself, the nurses and some of the patients were standing by the bed of one of the patients, who was just breathing his last, one of them, who lisps, broke the solemn silence by saying with a sigh, and slowly and solemnly,—

" Heth justh gone up the thspout ! "

MARCH 16.

All are better in my ward; except one who was brought in some four days since, and who will probably not survive twenty-four hours. Nineteen out of the thirty-nine at present here, are to have furloughs to their own States. Have been waiting some time for their transferal before taking a furlough.

Yesterday, went over to Louisville, on the ferry-boat which is so strangely named " John Shallcross," the name of one of the owners. " Sue Mundy," alias Jerome Clark, a noted guerrilla was executed. We heard the drum and saw people going to the terrible sight-seeing.

Received Government tickets for furlough.

Day before yesterday listened to an interesting lecture in our chapel by Dr. J. S. Newbery, Sanitary agent of Louisville. Subject—California.

Powell, of Adams County, Indiana, died last night. Write and send lock of hair as usual, to his wife. Also, for Samuel B. Sefton who died in Ward 7, formerly from this ward. For the former, two days since, I read the *first* letter he had ever received in his life. He is twenty-six years of age, has a wife and two children.

The paymaster has been here and some of the boys have had too much of him. A quarrel to-night in my ward, and a fight in Ward 6, in which two men were shot. The guard-house is full. Pity the sutler could not be tied up by his thumbs in the place of one who has had too much of his *beer* (?) Coaxed away a jar of brandy peaches from one of my patients and substituted *canned* peaches. Had I not, some three or four of the patients would have been too tipsey for their own self-possession in a short time and seen the guard-house before morning.

31

MARCH 19.

A sad day for us all. Dr. C. our ward surgeon, received orders to report for duty to Nashville. The patients are all very sad, and he feels the parting also. We improvised a lit tle oyster supper after the table was cleared of the full diet, for the ward-master and the old nurses. A torn table cloth from that Sanitary-box of rags, was cut in two for the narrow pine board, and looked quite like civilized life. We had oysters, a can of peaches, fresh butter and crackers, purchased of the sutler, and had a sadly lengthened meal. He goes in the morning.

MARCH 22.

One of the young nurses in the ward, who told me yesterday morning that he "had to get tipsey for the first time yet in his life," was last night unable to walk straight, and distinguished himself by talking loud, enacting the braggadoci﹥ by that, and by kicking over spittoons. He was coaxed off to bed at the tents, to prevent his being taken to the guardhouse. This morning, the boys were joking him as I entered the ward, when he said they "all talked as if he was drunk, when he wasn't at all. But Charley told him that "a man must be pretty far gone when he would feel his own pulse to see whether he was dead or not," which he had confessed doing the day before when he woke up at the tents, "because he had felt so strangely." Jehu confesses that though never in the guard-house for drinking, yet he served considerable of his time in one while at Memphis, but says that it was all on account of the pigs, turkeys and chickens in Tennessee, that they would bite a fellow so that he was *obliged* to kill them in self defence."

"That's so," says Willie B. most seriously, and with an ominous shake of his head, upon which he wears a famous white cap of my making, to hide his shaven crown, and snapping that one keen eye of his, " I declare if them Tennessee pigs and chickens don't beat everything. I tell you a fellow has to stand on his guard there, or they'd eat him up !"

Willie, by the by, says he has " lost the last cap I made for him to wear o'nights, and he suspects the executive officer must have confiscated it, when he was round on inspection, last Sunday morning."

A humorous patient, who professed to rejoice in the initials of D. G. W. G. II. A., and the corresponding short name of Don Garabaldi Ulysses Gabriel Hall Adams, has been tantalizing my pencil, one moment by very interesting recitals of hair-breadth escapes as a spy and among guerrillas, and the next by assuring me with equal gravity that he is first cousin to Gen. Grant, second to Sherman and third to Garibaldi, or something else equally incredible.

CHAPTER XI.

JEFFERSON HOSPITAL, March 27, 1865.

" You can charge it to the Sanitary ! "

Now it came to pass these words were spoken upon this wise : the sanitary carriage had started out from the hospital when we saw two men—the elder carrying a portmanteau and evidently the father—the other, a pale, emaciated invalid, who with feeble and uncertain steps was following. The carriage halted. " Wont you ride, and where do you wish to go ? " were the queries. These elicited the facts that the son was wounded, had two ribs broken, had had gangrene, had obtained a special transfer to Camp Denison near his home, and his father had come for him. After some hesitation the sick boy entered the carriage and was taken to the ferry. As his father helped him out he inquired, somewhat nervously, probably rating the fare in proportion to the easy cushions,

" What is the bill ? "

" You can charge it to the Sanitary," said the little lady as she wheeled the ponies.

Yes, Northern friends, if that dear one of yours who has been sick or wounded and in hospital, is ever nursed back to health and life, and restored to your arms again—bearing honorable scars it may be, or the loss of an arm or limb, but your darling and a hero nevertheless—if the truth were known, you could often " charge it to the Sanitary." And even he might not have known it himself. We deal out in such bounteous measure the gifts of the good genius, that often we do not think to say to the recipient, " This is a Sanitary

carriage you take your first ride for months in this morning,'
or " this is a Sanitary sling, shirt or handkerchief, pad, pillow
or crutch." And the fresh egg, lemon, orange, apple-butter,
blackberry jam, canned peaches, berries, or the cordial, jelly,
wine or green tea, may not often come with the word Sanitary ; but if he says it " makes him think of home," we often
tell him it came from there, if home means North, East or
West. Some go down even to the gates of death and are
won back by these agencies in the hands of a loving father,
without knowing it, while still others are deeply sensitive of
both the presence and the shield from death.

" I know that dried beef saved my life," said a sufferer in
the gangrene ward, " I could not, positively, eat a mouthful of
anything for days, till Mrs. B. cooked me some of that. Then
she brought me some every day till my appetite came for
other things." Another pale, emaciated man—a Frenchman,
in the same ward, said the same thing of potato soup and
green tea, who I found had eaten nothing previously for four
days. Said a German, in the same place, under whose arm I
put a soft cotton pad, " Oh ! I wouldn't take ten dollars for
that pad, it is *so nice*, and my arm was getting so bed sore
lying on these hard husks !"

This same pad, by the by, had come from some aid society
filled with rags—not a very soft cushion for a wound to lie
on. I had thrown the rags away and substituted cotton.
Pads will do very well as props, filled with straw or hay, if
there are cotton ones to lay above, next the limb, but rags had
better be sold as such, rather than pay transportation thereon.
Of course I shall not be understood as referring to white
rags which are large enough to dress wounds, those never
come amiss. But do not mark boxes or barrels containing
those as " bandages," the latter being much more plentiful
than the former.

32

"I do believe I would be willing to give ten dollars for a a feather pillow to lay my head on to-night," said that young hero "Willie" who had run away from bloodhounds in Alabama, and whose shaven head was throbbing with the disfiguring crysipelas. He had the pillow, and it came without money and without price, a gift from some noble, unknown Northern lady.

"God bless them," have I mentally ejaculated scores of times upon such occasions, or when the jams, the pickles, peaches, berries, or other delicacy, was the only thing which the palate would not refuse, and by which could be coaxed back the appetite. Oh, if they could only know and see, as we do, the lives saved, or the hours lengthened, comforted and cheered, they would not let the aid society run down, and the cucumbers and tomatos become the victim of King Frost or procrastination. They would not spend so much time in talk over their tea, of this husband or that son, but would work, if peradventure the fortunes of war should throw that son or husband into the channel of this bounty. They would not cease their labor of love because this one or that one had returned with the story that he had been in hospitals, and never had anything from the Sanitary, with his wise opinion that the surgeons and nurses were the only recipients. If he were closely questioned or his clothes examined, the scrutiny might betray the fact, that stationery and reading matter at least had come to him by that source, if he was not even then using a sling, crutch, handkerchief or shirt bearing the mark of some aid society. Or the complainant may have been situated somewhat as those of the 8th and 18th Indiana regiments, which have been constantly travelling, and never accessible to Sanitary stores for four years, until now, at Savannah, they have received a perfect God-send, and have occasion to use, as we are informed, the biggest word a soldier can say when overjoyed, "Bully!" Or he may have been honest in his convictions and truthful in his statements, for he might not have been needy.

Would you have your son, who perhaps was only suffering for a few days through exhaustion and exposure, or a slight flesh wound, who needed only rest, and whose appetite was good, eat the berries which might save the life of one in whose veins the fever had rioted for months, because *you* sent them. Or, if he had money with which to buy, would you have those warm socks, or flannels given to him, while that shivering, rheumatic patient, or the one convalescing from fever, and who has not received a dollar of pay in six months, went without because *you* sent them? Of course, you would not. You would prefer that your son should go through his whole life, with his present excellent opinion of surgeons and nurses.

The Sanitary stores are not so inexhaustable, nor the army so few in numbers that the former professes to supply fruit-cake and waffles to every mother's son who chances to stop a few days in a hospital.

Again, the world in a hospital is much like that outside. It has its share of grumblers and ungrateful ones, albeit there are those who cherish the idea that every soldier is one of nature's noblemen. Although there are many such who will meet you with a grateful smile in the morning, and the words that they are "getting better," while only close inquiry will reveal the fact that extreme pain has kept them awake all night, and banished peace by day, yet there are also libellers upon the character of noblemen. Among such are some who enter the army as substitutes, or volunteers *for the bounty*, who knew they were diseased to such an extent, that they would serve most of their time in a hospital.

Of this class will come limping up to the surgeon, one who

is grievously afflicted with "rheumatism," whining that he
has "been in the service three months, and hasn't had a furlough yet!" One such wished me to intercede for him. He
"shouldn't care so much about going home, but his wife wasn't
expected to live." So a letter just received had informed
him. I read it and found it written by the lady herself, while
inquiry revealed the fact that she "wasn't expected to live"
when he volunteered, but that the town was offering a high
bounty just then.

Another, who had "aphonia," when the surgeon was near,
but who could speak loud enough when complaining of his
food, or begging me for canned fruit, and because he did not get
it sneered at the idea of sick soldiers ever getting Sanitary
stores. When he found there were those who could read
him, he concluded he might as well get up, and he soon was
sent to his proper place, "the front."

Another, who had been nursed up from the grave's mouth
with delicacies and flannels, sold the latter, before going home
on furlough. And still another shot one of his own fingers
off, in the battle at Nashville, to get off the field. He was
the recipient of much sympathy, on account of his hand, being threatened with gangrene and amputation; but had the
facts been known before he was transferred to Louisville, I
verily believe I should have been tempted to try what my
conversational powers might do toward quizzing him to death.

" This mutton is poor substitute for chicken," said a grumbler in one of the wards, as the lady carried his dinner to
him.

" Well, yes," she replied pleasantly, " but I believe you
have been a substitute for a well man, for some eight months,
have you not?" He good-humoredly confessed that he had,
and had received a thousand dollars for the substitution. But
such, if really sick, must be cared for, as well as the more

worthy. Indeed, even a Confederate should not suffer at our hands, notwithstanding the brutality shown our loyal boys at Andersonville, and Libby.

It has truly been such a great pleasure to distribute that nice supply of stores brought to me by Rev. D. P. Livermore. The stock of flannel and canned fruit is nearly gone, as the distribution has not been confined to my own ward, but I believe it has not been misapplied. Blessings on the several donors and the agencies through which they came. Let us not be weary in well-doing, while the war lasts, for the end cometh. Some must "repair the breach, and build up the waste places afterward," but there will be no fitter time in which to make one of that number, who, when the All-Father cometh to make up his jewels, may hear the blessing: "Inasmuch as ye did it unto one of the least of these, ye did it unto me."

INDIANAPOLIS, IND., April 22, 1866.

A day of mourning in the calendar of a nation." A great grief sits sobbing upon a nation's heart, for *Lincoln is assassinated!*

This morning, while Miss C. and I were dressing, Miss T. rushed into the room with blanched face and exclaimed with grieved voice, " Oh ! girls have you heard the dreadful news ?" She knew we had not if she had thought, for she had left the room but a moment before, and she continued " Lincoln is assassinated, and his son and Secretary Seward." So the telegram at first was interpreted. It was a terrible shock, and I felt how almost as nothing in comparison would be the result of the death of the dearest friend or relative I had, and believe then I could have given my own life could it have restored his life for the country. Little has been done by any of us to-day. Toward evening we three visited the hospital and saw all along the way, wealth and poverty, the mansion and the hovel displaying symbols of grief. The man of talent remembers that the eloquent man, the counsellor, has fallen, while the man who returns at night with his daily wages thinks with sorrow of that one, who from greater poverty than his own, has come to be the mourned of a nation.

As for the lesson of this deed, I cannot soothe myself as do some, with the thought that Lincoln had done all he could in this war, that his heart was so tender he could not deal justly with traitors, and that his mantle has fallen upon one, who is "sufficient for these things." Instead, this climax to rebel atrocity, approached only by the starving of our brave boys in prison, seems to me to call for stern justice to be meted out. All the blood and treasure of the last four years demands it, and now *this brother's blood* crieth to us from the ground. Will it be heard, or will rebel traitors take their seats in congress to make laws for those who have shouldered their rifles in defence of law and against traitors and assassins? We shall see!

JEFFERSON, HOSPITAL, May 3.

" A hospital is no place to form attachments," said one lady in this hospital to another. The former had surprised the latter in a sudden flood of tears, in the pantry of Ward 1. The occasion was the arrival of that order for the " kitchen man, and chief wound-dresser," of said ward to report to their own State, New Jersey.

Perhaps it is not wise to form attachments, but if they grow themselves, as between a mother and sick child, with every cry of pain, or bestowal of attention, what is one to do about it? It is quite inconvenient sometimes, I admit. But I would like to see one who is created with that troublesome thing, a heart, and who takes care of patients, from the time they are brought in just from the front, looking more like wild brigands from the mountains, or Indian trappers from the frontier, so far as hair or whiskers are concerned, but acting more like babies, or, it may be, like very sick but stout-hearted heroes, but who after they are bathed, provided with clean clothes and bed, and the superfluous hair and whiskers removed, turn out respectable-looking, civilized beings, up to the time when the departed appetite is coaxed back, and when by pleasant conversation, letter writing and reading the relaxed nerves recover their tone and grim death is fairly beaten back, who at first had a mortgage upon them,—let such an one have a care for the feet on scrubbing days, when they are able to sit up, and muffle them for a ride in the Sanitary carriage, to get a fresh breath and sight out of doors, the first for months, and just when she knows that a sudden relapse might take them away, to have an order come for a transfer to Quincy, Keokuk, or Washington, and she would probably feel the "attachment," if she possessed that troublesome thing, a heart. And, by the by, I happen to know that this same Miss B. who gave the caution, has in possession a pretty good-sized article of the same kind herself.

It happens that my large family of boys, being under the guardianship of their Uncle Sam, are liable at any time to be torn from my maternal oversight, to go either to hospitals elsewhere, or to their own regiments. I derive, however a sort of savage pleasure, from the evident and acknowledged fact that they "hate it," as the Egyptians say, as badly as I do.

And the separation may be equally felt under other circumstances. This is the case in the transfer of those whose presence seems indispensable to the good of the patients. To wounded men, who have learned to have confidence in the skill, care and tenderness of a wound-dresser, it seems little

less than cruelty to send him away, and substitute one new
and inexperienced, especially when a little less care than usual
may inoculate the wound with erysipelas, or gangrene.

Or it may be one to whom we have all looked up as a
counsellor, whose rich humor and dry jokes were a never-fail-
ing fund of enjoyment to the patients, and who was a walking
enclyclopedia for their benefit and my own, and who with
such an influence in the ward, treated me with the greatest
respect before all, with such fatherly forethought, and whose
child-like innocence was a constant reproof to any thought of
wrong.

Not a surgeon, ward-master, or nurse remains of those who
were here six months ago, while some of the nurses are in
hospitals elsewhere, and most in the position of patients.

Well, the work and care for the sick boys, with this tearing
of the heart-strings every few days, didn't seem to have a very
beneficial influence upon health and nerves. "Pale, care-
worn and thin," was the verdict of others, while myself only
knew the extent of the malady and the need of rest, when I
found that twice I had actually cried like a child, because loud
talking in the night and building fires before reveille in the
morning, had waked me. Not having for some time been
able to sit up all day, though attending to my duties in the
ward, and as transfers had taken nearly all the sick men,
chancing to leave convalescents, I decided to run away for a
little time, where I could rest, eat and sleep. Dreading the
long jaunt north to my friends, I accepted the urgent invitation from a lady friend and co-worker, to visit her people,
near Pendleton, Ind., and procured a furlough of twenty days.
No sick soldier could have been more thoughtfully cared for,
in the home of Mr. Neal Hardy, than was I. This neighborhood itself has abounded in works of charity to our sick
soldiers during the war, and many boxes and barrels packed

by the hands of Mrs. II. have gladdened many an invalid soldier. I had there good nourishing food, of which I was greatly in need, for if, as is reported North, the surgeons, stewards and nurses eat all the sanitary stores, our "ladies' mess," has certainly failed to obtain its share of the plunder. But the nourishing food I found on my furlough, with sleep, freedom from care, and genial companionship, when I wished to avail myself of it, for the time being, wrought a cure.

Upon my return, had expected the patients would be glad to see me, but had scarcely looked for so warm a welcome as was received. The next eve, we had a nice supper for the patients of Ward 1. During my furlough, the good friends donated a box of eatables for the use of my ward and for that of the lady at whose house I had been visiting. Just before supper, Miss H. and myself surprised the boys by carrying several articles into our pantry, and preparing for the table goodies which had not appeared on the printed list as "full diet." An old table cloth and sheet which came from Chicago in that box of sanitary rags, was torn into strips and placed on our three long narrow tables. Three or four were watching me.

"Boys, we're going to try if we cannot make you think you're at home to-night."

"Well, I declare it'll be the first time I've sat down to a table-cloth in eighteen months," said one. "And the first time I have, in three years," said another. "Its the first time in nearly four years for me," said a third.

Then the plates were turned down, and the food put on other plates and in bowls, instead of being dealt out on each plate, as is usual here before setting down. The Chaplain's orderly was present, Mr. Bullard, of Illinois, who was formerly a patient in our ward. After a blessing was asked, the food was passed, but at first every thing was so strange that all were glum and silent.

33

It was evident we were to have a solemn time, only to be remembered by them as one in which there was "a putting on of too much style for comfort," as they would have expressed it, so they were told the intention was to make them feel at home, and if they were there they would surely talk, and as we had plenty of time, we would try to have a social time as well as a good supper. Whereupon our theological student, who has since left to receive a 1st Lieutenant's shoulder-straps, timidly remarked, that, " The trouble is, most of the boys think they're out taking tea somewhere, and *durstn't* say anything." Then " Bart," as he is familiarly called, looked around, and said hastily, as if grudging the time occupied in speaking, and with his half-lisping accent, " Boys, I'm intending to say something of considerable importance pretty soon, but I'm too busy with my supper just now."

This helped break the ice, for the boys feel bound to laugh whenever Bart says anything. Soon, leaving the room for a forgotten article, I charged them not to speak aloud until my return, and appointed a monitor. Upon my entrance, our " little artist," Hugo, in that tone of complaint used by children to their teachers, in the school-room, said that, " Bart Smith commenced it, for he said he wished the lady nurses would go away on furlough every week." Of course it was necessary to rebuke him for wishing our absence, when in a tone of conciliation he informed us that he " knew'" he "said that, but had said also that he wished them to come back the next day."

This, with the entrance of Miss Buckel and another lady, who contributed to the pleasantry, made them completely at home, and every one seemed thoroughly to enjoy the supper. The regular diet for the meal, which was sent, was merely " bread, stewed apples, and tea." The apples were saved to add to their breakfast, and apple-butter supplied in their place.

In addition, we had fresh butter, horse-radish, berries, cake and chickens, with sugar for tea.

This much for the supper. May it not have been a link in that chain of "attachment," by which many an old soldier in the years to come shall feel bound to the large family of brothers and one sister in Ward 1 ?

MAY 8.

Nations are divided, thrones totter, confederates are captured and hospitals are broken up! Consequently, everybody is on the tiptoe of expectation, or in the slough of despondency, over the coming separation. It is somewhat curious to note the changes and effects of the order commanding the discharge of all soldiers except veterans or those under medical treatment. Some of the former try to pass themselves off as later recruits, to the infinite disgust of the official who questions them; while others immediately "throw physic to the dogs," to prove that they are not under treatment. Every possible rumor is afloat. It is even whispered by soldiers that the "dignitaries" begin to have an inkling of the fact that it will not be long before they will have no more authority than "high privates," and have relaxed a trifle from their dignity—I am, however, not responsible for the truth of this, and learned sometime since to my own edification, that women are not the only gossips.

Some of the wards here have been closed, and the patients transferred to others. Consequently, some of the lady nurses are as wholly lost and inconsolable as a mother-hen, who, by some terrible calamity, has been cruelly deprived of every darling chicken. We, who have thus suffered, do not much care whether the world stands any longer or not, our housekeeping has been cruelly broken up, and we should doubtless throw ourselves into the Ohio and thus drown our sorrows, were it not that said river is altogether too muddy just now.

The effects of said "order" upon the soldiers, as appearing in my own ward, are with little variation, no doubt, the same as throughout the hospital.

Our patients had been tranferred to Ward 2, to give opportunity for floor-planing and white-washing. But all who could walk, made frequent "visits home," indeed, extending to visitations, and two boys, Willie and our "little artist," could not sleep a wink the first two nights, and were allowed the privilege of sleeping at home a few times. One morning, while our floor was in process of planing, upon entering the ward, I found the tools thrown aside, and all seated in a group, reading and discussing a Louisville daily. All were jubilant, and eager to tell me the news.

"Can't work," said one, "*too* much good news for one day! Johnson is taken, and we hospital bummers are to be sent home," said another.

"It's wicked to work any more, an order has come for all Government work to be stopped," said a third. "Goldsmith's Corps will soon be on the wing."

"Boys let's pile the shavings in the middle of the floor and have a bonfire," suggested the one who bears the name of "Gen. Grant."

By the way, very few men in camp, or hospital are called by their names. Instead, we have "Gen. Thomas," "Cavalry," "Artillery," "Michigan," "Connecticut," "Georgia," "Longstreet," "Infant," "Lengthy," and "Bantie," the last four named from their height, or want of the same. Other States have their representatives by name, while occasionally so suggestive a title as that of "The Spread Eagle," is overheard while the owner is at a safe distance.

MAY 15.

Our flower gardens are now absorbing the attention of the ladies and convalescents. The arrangement of the hospital is such that there is sufficient space for one between each ward. In this respect, at least, this must be superior to the "Chestnut Hill," hospital, of Philadelphia, as these wards, radiating like the spokes of a wheel from a circular corridor, not only permit the addition, for the comfort and cheer of the invalids, of fresh flowers and the sight of a green-sward close to their windows, but also superior ventillation. It is to be hoped that so long as a hospital is needed for sick soldiers, this will be taken for no other purpose. And when no longer needed for the sick, what place could be found more suitable for a "Soldier's Home," for the loyal boys of Indiana?

The flower gardens were, by Major Goldsmith, given over to the superintendence of Miss Buckel, who has charge of the ladies here. She preferred that those of each ward should originate and carry into effect their own plans, while she procured shrubbery, seeds, and plants for all. None have refused to whom she has applied, and most have responded liberally. One gentleman from Chicago, donated $25 worth of rakes and trowels.

The friends at some Aid Society, I think in Ohio, contributed a very large and choice collection of seeds, at the request of a patient, Chas. Erickson, in our own ward. These were shared by all the ward gardens.

I am sometimes amused at the difference it makes with the patients as to who asks them to work. Most of them say they are not going to stay long enough to see the gardens after they are finished, and they don't care to work for nothing. Those in our ward often refuse to work upon being requested by the ward-master, who will nevertheless work nearly all day. Others run away from work elsewhere, as on the chapel,

34

or the garden around it, both in care of Chaplain Olmstead, and work in our own garden. This morning we were mustering our little force, and I was in the garden when the ward-master came and said that "Jehu had sworn he wouldn't work a bit to-day, but that as he was as well able as the rest, and as he had received orders to put all in the guard-house who would not work, he should certainly send him there, unless he changed his mind."

After cautioning him not to let J. know that he had told me this, I stepped to the window near which I had just left him as I passed through the ward, tapped on the pane, and said, "Jehu, do you suppose I can get you to help me trim this sod around the beds?"

"Yes, of course you can," he exclaimed energetically, and springing up he ran to the ward-master for the keys to the pantry to get each of us a knife for the purpose, and then jumped from the corridor window and helped me until the work was finished.

Then Miss B. came to say that we could have carts and mules to draw sod, if I could find drivers and sod cutters.

"Three have volunteered," I said, wonder where I can find another?"

"Why, you can get me, if you want me," said Jehu, earnestly.

He went and worked well all the forenoon, but in the afternoon, word came that he had refused to do anything more, and was asleep under a tree. "Tell J. for me," was the word sent, "I wish to know if he won't please help us a little while longer, as we may not be able to get the carts tomorrow."

It was sufficient, six carts of sod were cut by him, and one other.

The work was at first somewhat delayed by the scarcity of tools, and since by heavy rains, but is steadily progressing.

Some gardens are finished, or nearly so. We have quite a variety in style, from that which bears the cognomen of " Methodist," or " Quaker," to the one which will contain only our " Star Spangled Banner," with its stripes in red and white flowers, blue for the ground of the corner, and shells for the stars. I *do* hope all this labor is not destined to be trampled under foot by those who do not so appreciate, or need it, as do sick men.

At first but few volunteered to do the work among the convalescents, but soon others became interested, and in some wards the excitement was such that some men even choose rainy days to go down town, rather than working ones. This interest, with the enthusiasm from the prospect of a speedy return home, has occasionally led to scenes and conversations really amusing. Sometimes a number of States have their representative by name, who work with a will to prove that " Massachusetts can do the most spading," or that " Michigan can't be beat at sodding." In our garden, they were one afternoon engaged in erecting a large mound in the centre. Some one of the members had tied a newspaper to the end of a pole and hoisted it.

" Boys, you are not working as you would, if really throwing up breastworks under the guns of the enemy," said one of the patients from an open window.

Just then it chanced that there was the report of fire-arms at no great distance. Two as suddenly reeled, one falling to his knees, but recovering his feet ran for the pole which had fallen, and planting it firmly, called out,

" Hurry up, boys, here's our flag of truce, and the enemy will respect it."

They then did work as though under the guns of the enemy, and the mound was soon finished.

But little has been done here for the forthcoming great

Sanitary Fair at Chicago, in consequence of this great work of the gardens, though much interest might easily be aroused and work done, if the ladies had not these to occupy head and hand. One life-like sketch of " The Trapper's Last Shot " is nearly finished for the fair, by " our little artist " Hugo. It might be interesting to the future purchaser could he know that since commencing the sketch the artist has had gangrene in one of his wounds, and has done much of it while sitting upon his bed, and when it was not prudent to exercise by walking. I shall envy the fortunate possessor of the picture, although the one from which the sketch is made is in my own possession.

MAY 24.

No sick men in my ward. It has been filled up with detailed men from the tents. Most of our former patients remain in Ward 2, and I assist Mrs. D. in care of them; besides which the only patient of Ward 8 is left to my care, as the lady has gone.

An order came sometime since for the discharge of all hospital attendants not absolutely needed. In pursuance of this order seven ladies have received their discharges, and the last except one of them go to-night. There are some fourteen remaining.

One of the " Willies," whose home is only eight miles distant, and who has tried for six months to get a furlough, has just returned from a second " French " furlough. Nobody missed him, who would report the same.

This morning four ladies went over to Louisville in an ambulance, the principal errand of two of us to purchase some little gift for Miss B. It was a beautiful wrought silver card case, which somebody had the pleasure of presenting at the tea-table this evening with the words:

"Miss B——, I have been requested by the ladies to present you with a small token of our esteem. In years to come, when thinking of your cares and duties here, may this little gift assure you that the responsibilities and difficulties of your position, and the faithfulness with which you have discharged them have not been unappreciated. The gift is small, but we believe you will value it nevertheless."

She had been met as she was leaving the table, and as the gift was in a morocco case she took it and said, "I accept it and will run away to see what it is." She soon returned and simply and naturally expressed thanks and admiration for the gift, and added playfully that she would "keep it as long as she lived and then will it to her grand-children."

MAY 29.

Presentations seem the order of the times. Quite a number of the ladies have been so honored by the patients of their wards. This afternoon I took my sewing down to the ward and was soon surprised to see Mr. Davis come hobbling in on crutches from Ward 2, who had not been in the ward, and hardly off his bed for two months. Then several others who had been patients here suddenly dropped in, as well as some others of Ward 2, and when Bart and Hugo came in on their crutches, I thought it rather a queer coincidence that they should all happen in so soon after I had entered, but supposed the beauty of the day and the desire for one more of those pleasant chats, which were so soon to be broken up, were the causes. But in a moment more it was explained by the entrance of the young man called "Kentucky"—Wm. Garrett —bearing a box from which he took a nice photograph album, of a size to hold one hundred pictures, and a ring, with my name engraven upon each, and which he presented with a few appropriate words, and with an easy, natural manner. They were assured that the recipient was not much accustomed to speech-making, but that no gift could have been more acceptable than the album, especially when it should contain the faces of the donors. The ring was found to be a perfect fit, and some wonder was expressed, but all were perfectly innocent and nobody knew anything more about it than parents do when that wonderful genius of children, Santa Claus, is making his annual visit. Then I chanced to recollect that Miss B. had tried my ring on some time before, and some whispering had occurred at the door upon the taking of a box from her room, when I chanced to be there two days since.

JUNE 1.

Next day after the last date I went down town early in the morning, purchased strawberries and made a strawberry short-cake for tea. It was very nice, greatly complimented, and by some who had never seen one before, but who were going to have their "wives make one as soon as they got home." In addition we had green tea with milk and sugar— a great treat here—stewed prunes, cooked tomatos, very nice dried beef and cookies. All old patients of our ward were invited throughout the hospital, with the ward-master and nurses of Ward 2. Some few days since we also had a pleasant little time with strawberries and cream provided by Mrs. Dixon in Ward 2.

The hospital is being thinned out quite fast, but much too slow for the patience of most of the soldiers. Nearly every day I am called to part with some one or more of the old patients.

It is, as was predicted, about those several hundred volumes being kept for a reading-room to be finished and fitted up, and the soldiers deprived of their use. Thousands have passed through the hospital this past eight months, and those books have been boxed up which might have given occupation, relief from home-sickness, to say nothing of mental, moral or spiritual improvement to invalid soldiers. What books Chaplain Fitch had in charge have been freely distributed, but they were few in number and of very little variety. There has been some pressure brought to bear of late by some ladies and others, and by the offer of Rev. H. F. Miller, Agent of Universalist Army Mission, to bring his library, which may result in the unboxing of the books at the eleventh hour.

A very pleasant reading-room has just been fitted up by the ladies, in one of the vacant wards, and Chaplain Fitch has procured the loan of two libraries from the Christian Commission. This is very pleasant and is greatly enjoyed by the patients who make it a resort for reading, writing, chatting, or the amusement of checkers, chess and backgammon. Pictures adorn the walls, there are plants in blossom, and each day is a beautiful bouquet contributed from one of the ward gardens for the hanging flower-vase. All enjoy this very much; but it only reminds some of us of what might have been all winter, had the one who had the chapel and library in his hands have so willed it. I understand that several thousand dollars were placed in the Chaplain's hands by the Sanitary or Christian Commission for the purpose, and which has been of no comparative benefit.

Walking along the corridor one rainy day of late I picked up a wee little book with the following revery, entitled

A RAINY DAY IN CAMP.

It's a cheerless, lonesome evening,
When the soaking sodden ground
Will not echo to the foot-fall
Of the sentinel's dull round.

God's blue star-spangled banner
  To-night is not unfurled;
Surely *He* has not deserted
  This weary, warring world.

I peer into the darkness,
  And the crowding fancies come;
The night-wind blowing Northward,
  Carries all my heart toward home.

For I 'listed in this army,
  Not exactly to my mind;
But my country called for helpers,
  And I couldn't stay behind.

So I've had a sight of drilling,
  And have roughed it many ways,
And Death has nearly had me,
  Yet I think the service pays.

It's a blessed sort of feeling,
  That though you live or die,
You have helped your bleeding country,
  And fought right loyaly.

But I can't help thinking sometimes,
  When a wet day's leisure comes,
That I hear the old home voices,
  Talking louder than the drums.

And the far, familiar faces
  Peep in at the tent door,
And the little children's footsteps
  Go pit-pat on the floor.

I can't help thinking, somehow,
  Of what the Parson reads,
All about that other warfare,
  Which every true man leads.

And wife, soft-hearted creature,
  Seems a saying in my ear,

"I'd rather have you in *those* ranks,
Than to see you Brigadier."

I call myself a brave one,
But in my heart I lie!
For my country and her honor
I am fiercely free to die;

But when the Lord who bought me
Asks for my service here,
To "fight the good fight" faithfully,
I'm skulking in the rear.

And yet I know this Captain
All love and care to be:
He would never get impatient
With a raw recruit like me.

And I know He'd not forget me
When the Day of Peace appears;
I should share with him the victory
Of all His volunteers.

And it's kind of cheerful, thinking,
Beside the dull tent fire,
About that big promotion,
When He says, "Come up higher!"

And though it's dismal, rainy,
Even now, with thoughts of Him,
Camp life looks extra cheery,
And death a deal less grim.

For I seem to see him waiting,
Where a gathered Heaven greets
A great victorious army,
Surging up the golden streets;

And I hear him read the roll-call,
And my heart is all aflame,
When the dear Recording Angel
Writes down my happy name!

35

But my fire is dead white ashes,
  And the tent is chilling cold,
And I'm playing *win the battle*,
  When I've never been enrolled.

In Thine army vast receive me,
  Thou Saviour of the world !
And I'll follow wheresoever
  Thy banner is unfurled.

Oh, give me zeal and courage,
  My heart and life renew,
That I firmly to my signet
  May set that Thou art true.

To reach the Eternal City,
  I'll brave Death's sullen flood,
My Saviour crossed before me,
  I'll triumph through his blood !

JUNE 13.

Many things of interest occur which I have neglected to note. The truth is, am getting ill again. Have been so sorry to see former symptoms all coming back, as it is a sign that I must leave the hospital. But of late have really not been able to sit up all day, and am kept awake at night by a cough. My lungs have been examined by Miss B——, who pronounces the left affected, and prescribes " Hygeine and California." Think I shall take a dose of both. Have made application for discharge to be given in one week.

Four other ladies and myself have of late been filling out discharges at headquarters. Several convalescents are detailed also, and with the clerks proper we are making out the mustering-out rolls of from fifty to sixty men each day. There are eight of these papers filled out for each man, besides the discharge proper.

One day not long since, while busied with sewing in my

ward, several were relating incidents of their experience, two or three of which I will mention. Mr. J. of the gangrene ward said, that once when with his regiment, Wolford's Cavalry, and near the line of Virginia and Tennessee, a woman in front of her house watched them for some time, and then asked. " Whar be you'ns from anyhow ?"

" From Ohio," was the reply.

" La, now," that 'Hio must be a mighty big town to have so many men in it, is it in Tennessee, or Cincinnati ?"

At another place they formed in line of battle along the street, and in so doing sadly discomposed the lye apparatus of another woman. She was indignant.

"Thars that Wolford's men come down yere, a creeter-backed, to fight weuns, and they formed a streak o' fight, and knocked over my new ash hopper what cost me ten dollars an' a half, and never paid me a cent !"

Another related the following incident. It was after a battle on the Mississippi River that a captain on one of the river steamers offered to carry free of charge a certain regiment who were engaged in the battle. At each trip many presented themselves as members of that regiment. At last one stepping on board reported himself as a member of the same, when the captain asked what office he held.

" Not any," said the soldier, " I'm a high private."

" Give us your hand," said the captain, "glad to make your acquaintance, sir ; for you are the first private I have met from the regiment—have passed up over two thousand, but they were all officers."

To add my mite to the story-telling I related an incident which occurred while I was in Nashville, and which I heard related by the young minister concerned, and at the time of its occurrence. He had been engaged to perform the marriage ceremony at a certain hour for a couple of Refugees at the Refugee Home. It was but a little distance, and as he started out, a few moments only past the appointed time, in company with another delegate, he saw the bridal pair with another couple coming. They met upon the lawn, and the young clergyman told them he would perform the marriage ceremony there if they wished. No objection was made, and it was accordingly done, when the minister wished the bride " happiness in the new relation," and she *wished him the same.* As they were about leaving the gentleman who accompanied the clergyman offered to shake hands with the young lady who came as bridesmaid, but suddenly withdrawing her hand with a frightened voice and manner she exclaimed,

" Oh ! I don't want to be married—I ain't ready yet !"

During this conversation the door opened and Revs. Fitch and Miller entered, and the former said, with his characteristic gallantry,

" This is a *vacant* ward, Mr. Miller, there's nobody in it, but you see what a crowd somebody always has around her," and then followed more nonsense upon the same subject, when he was informed that we had been listening to large stories from the soldiers, but had scarcely expected one from a chaplain.

SABBATH, JUNE 18.

My last day at the hospital. I leave to-morrow. Early this morning picked the last bouquet from our garden to place in the ward, and pulled the first mess of radishes therefrom and prepared for the table of Ward 2 for dinner. Besides flowers, these, with lettuces, were the only vegetables planted.

Thought has been busied with retrospection to-day, and with the subject of woman's influence in a hospital. And notwithstanding that there is much feeling upon the subject of her real or imagined interference with professional duties, yet there are very many wise and noble surgeons in the service who rightly appreciate woman's influence in a hospital, and have assisted her in every noble word and work. And a pure, true woman is amply repaid for working her way quietly and kindly against opposite influences, as she may feel assured that her efforts are blessed to the sick boys in her care. She is amply repaid if at the last she may so overcome the prejudices of a physician as to hear him say what was said to one of the number :—

" You have been a blessing to the patients and a help to me—have attended to your own duties as nurse without interfering with those of mine as physician. And there are those whose lives are due to your care. Some were very low with nervous prostration and nostalgia—another name for home-sickness — and your conversation and attention has aroused, cheered, strengthened and saved them."

Or if she may hear from one and another patient, as the same one has when bidding them good-bye for the last time, such words as these :—

" I shall never come down again as I did here to what I thought was my death-bed, with so little preparation. I'm going to make it a first business of my life to learn how to live, that I may not be afraid to die, and if ever I am a better man it will be due to your influence and your counsels. May God forever bless you ! "

Or if one might have such a beautiful tribute to the worth of woman's presence among sick soldiers, as our friend Miss Miller, of Chicago, received on board the floating hospital called the Nashville, near Vicksburg. There had been no white woman on the boat previous to her arrival. One afternoon, as she stepped into one of the wards for the first time, her ear caught an exclamation of surprise from the inmate of a bed not far distant, and turning in that direction she saw a sick soldier, with hands clasped and the great tears absolutely raining over his face, as he gratefully exclaimed :

"Thank God!—I can die easier now since I have seen a woman's face once more."

And despite the multiform abuses which have stained the records through every department, during this great rebellion, there has been wrought out a greater good and higher destiny for mankind, than we may well realize, and the former sink into insignificance in the majesty of its glorious presence. Like the poet, I

> " Have seen it in the watchfires
>   Of an hundred circling camps ; "

like him, have

> " Read it in a fiery gospel,
>   Writ in burnished rows of steel,
>   That God is marching on."

And we know, though wrong and oppression do exist in high places, yet it was not in vain that

> " They went forth to die!
> Unnamed, unnumbered, like the desert sand,
> Blown to build up a bulwark round some land,
> To stay the sea of wrong that vainly raves,
> Forever, on a shore of patriot graves,
>   That they went forth to die;"

Neither will it have been in vain to all future ages, that

" Ye went forth to save
The precious offerings, like the patriarch's, given
On high Moriah in the faith of Heaven,
To stay the knife ere yet its point be hurled
Through hearts which hold the promise of the world,
That ye went forth to save!"

www.ingramcontent.com/pod-product-compliance
Lightning Source LLC
Chambersburg PA
CBHW021704210326
41599CB00013B/1508